**Kinds of House Carpentry**

**Kinds and Grades of Lumber**

**Hand Tools and Power Machm...**

**Types of Joints**

**House Layout and Form Construction**

**Framing—Floors, Walls, Stairs, Ceiling and Roof**

**Exterior Finish—Cornice work, Windows, Doors**

**Interior Finish**

**Hardware Installation**

**J. DOUGLAS WILSON** has been active in this field for more than 30 years as a practical carpenter, a teacher, and author of books and articles on the subject. His latest professional connection was Supervisor of Vocational Curricula, Los Angeles City Schools.

# Practical
# House
# Carpentry

## Simplified Methods For Building

### J. Douglas Wilson

McGRAW-HILL BOOK COMPANY, INC.

New York   Toronto   London

PRACTICAL HOUSE CARPENTRY

# Preface

The great increase in the population of the United States during the last two decades, plus the accompanying demand for more and better housing and other facilities, has meant much to the building industry. Practically every section of the country has felt the effects of tremendous and increasing building programs. This is evidenced by the great number of new homes, stores, churches, industrial buildings, and public schools, as well as by much in the way of remodeling.

Such an increase in building requires that there be more building-trades carpenters. Thousands of young men, therefore, including veterans returning from recent wars, have entered the carpentry trade as a means of livelihood. Many of them have been forced, however, to restrict themselves to an elementary type of carpentry work, as they did not have time to prepare themselves properly through the more orderly processes of apprenticeship training.

This book has been written to serve this large group of partially trained carpenters. It contains a wealth of how-to-do-it information, acquired through careful training and long experience, which should be useful to all who plan to carry on carpentry work.

The various phases of house construction, as described in Chapters 5 through 9, are presented in a logical sequence, following the order of job procedures. Chapter 5, The Foundation, covers house layout and form construction; Chapter 6, Framing, includes underpinning, floors, walls, ceilings, and roofs; Chapter 7, Exterior Finish, pertains to frames, cornices, siding, and roofing; Chapter 8, Interior Finish, describes trim, cabinets, windows, sash, doors, and hardwood floors; and Chapter 9, Finish Hardware, deals mainly with the installation of finish hardware.

This constructional order of presentation enables the learner to begin with the first step in house construction and to follow through, in his learning process, in exactly the same manner as he would on an actual job. This arrangement also permits an apprentice or journeyman carpenter or a do-it-yourself person to study the procedures necessary to do the specific job in which he is interested.

Hand tools have been classified according to the particular phase of carpentry, such as rough framing or interior-finish work, in which they are required. They are carefully illustrated, and the use of each tool is carefully explained and integrated with all manipulative instructions.

v

The basic joints common to most carpentry jobs have been described, and the construction procedures necessary to secure maximum strength are carefully explained. The best type of joint to use for various specific jobs is indicated in Chapter 4, Joints. A thorough knowledge of joints is constantly required when planning for and executing a piece of carpentry joinery.

The set of blueprints at the end of the book is included for use by instructors when discussing various construction problems.

The instructions included in this book, all of which are expressed in terms that will be clear to the learner, cover in step-by-step order the basic operations required to construct a house, from the driving of the first stake to turning the key in the front-door lock. These basic carpentry operations, which are applicable to any type of residential framing construction, are described in detail and accompanied by working illustrations, so that all necessary visualization is assured the young mechanic as he prepares himself to become an expert carpenter. Instructors of carpentry will find these descriptions adaptable to their instructional procedures and coordinated with acceptable national building specifications and building codes.

The entire discussion, as well as the how-to-do-it style of presentation, is based on the author's many years of experience as a carpenter, foreman, and public-school carpentry teacher; during his period as a teacher, a number of houses were constructed from the ground up with students performing all the carpentry jobs involved. A long period as a supervisor of carpentry instruction for the United States Army has provided the author with proof that the basic principles of house carpentry are the same, irrespective of geographical location in the United States. Hence the building methods given are sound and practical, and, if carefully followed, will result in a building that is level, plumb, straight, and square—the four basic characteristics of a well-constructed house.

It must be emphasized that building ordinances should be carefully studied and observed. They have been written to protect cities from poor construction, unsanitary plumbing installations, and fire hazards. Buildings which are free from sags and cracks due to improper construction are also longer lasting. The materials used in house construction must be selected on the basis of local custom and availability.

Progress in carpentry comes only from studying the many problems of the trade and by working daily at the trade to acquire the necessary tool skills and experience. A careful reading and study of this book will augment the carpenter-learner's technical training and thus help him acquire the status of master carpenter.

J. DOUGLAS WILSON

# Contents

# CHAPTER ONE

# Introduction

In this chapter the history of carpentry will be briefly scanned, two general kinds of house carpentry will be noted, building ordinances and codes will be discussed, and, finally, the presentation plan for the book will be outlined. The purpose of this chapter, as its name implies, is to introduce the reader to the subject and to prepare him for the following chapters.

## HOW CARPENTRY DEVELOPED

From time immemorial, shelter has been one of man's primal needs. The first and most ancient attempts to build shelters may have taken place when cave dwellers were forced to leave their caves and seek new hunting grounds. That probably brought about a serious problem, for they then had to devise some new means of protection from the elements. The types of shelters first erected depended, to a great extent, upon the materials at hand in the immediate environment.

In hot and arid regions, trees and stone may have been scarce or entirely unobtainable. Thus, shelters were built with ordinary mud as the sole material. Examples of desert mud shelters are still in existence. In climates where trees, leaves, and grasses were available, rude shelters were erected using branches for framework and leaves and grasses for roofing and sheathing. In still other localities, where stone was easy to get, man learned how to use that material to advantage in building shelters.

There is definite evidence that wood, wherever it was available, became the favorite material for building shelters. This fact is substantiated by the many ancient stone woodworking tools which have been uncovered at various places around the world. Such tools made possible the first real carpentry work. As time went on, man learned to mine ore and to work with metals. Probably some of the first metallic operations by fire were employed to fashion greatly improved woodworking tools. Through the centuries, tools were slowly improved until carpentry work was firmly established.

Carpentry, as a craft, is actually thousands of years old. The earliest carpentry craftsmen applied their skills to the making of many products for which wood was the principal material. In the beginning of the eighteenth century, carpenters made all the fine interior cabinetwork to be found in the great houses, castles, and churches in Europe. They also designed and made the beautiful furniture of that period. Present-day "period" furniture is still copied from the original designs of those early master carpenters.

*Carpentry in America.* When the first colonial people arrived in America, their initial concern had to do with the building of suitable shelters. Like the earlier cave men, they had left their original homes and were faced with the problem of building new ones. Fortunately, wood was available in great abundance and was easy to work with. Early American carpenters were experts at cutting trees and shaping logs. Contrary to popular opinion, their houses were not of the type made famous by Abraham Lincoln. The Lincoln type of cabin originated as people moved westward and did not have the time to fashion the types of houses they left in the East. Their aim was to build shelters as quickly as possible so that they would have more time to devote to farming. Most of the colonial houses along the East Coast, however, were fashioned from logs which had been squared and jointed. Such houses were very beautiful examples of the skill those master carpenters developed, even though their tools were limited and crude in comparison with present-day tools.

As more time went by, machinery was developed which could cut tree logs into lumber of all shapes and sizes. Carpentry tools also improved more and more rapidly. With the improvement of both machinery and tools, houses likewise improved and became much more standardized in architecture and construction.

Today the work of carpenters has become specialized to a high degree. In many instances, carpenters work on mass-housing projects and confine their efforts and experience to only a few of the general skills a master carpenter should be able to exercise. They often lack many of the all-round skills of the carpenter who works on "tailor-made" houses and who has the capability and capacity to perform all the operations required in building a complete house. The purpose of this book is to explain all commonly encountered house-construction carpentry jobs, thus enabling the reader, through study and practice, to do a proficient job of construction, as measured by the prerequisites of having the building plumb, square, and level and of using good joinery.

## KINDS OF HOUSE CARPENTRY

A master carpentry craftsman has the all-round, or general, ability to do any and all kinds of carpentry. In other words, a skilled carpenter is one who can do all the carpentry work involved in the ordinary types of American homes. In most cases, and for the purpose of this book, such general carpentry can be divided into two basic kinds.

*Rough Carpentry.* Figure 1 shows some rough framing typical of the kind encountered in the average house. The various joists, studs, rafters,

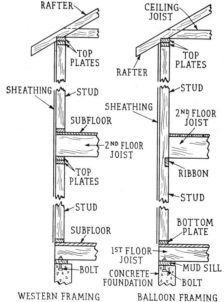

**Figure 1.** Examples of rough framing carpentry.

and sheathing, plus other structural details which are not visible after houses have been completed, constitute what is called *rough carpentry.*

The term *rough carpentry,* derived from the fact that the lumber used in this type of work is rough and not finish lumber, should not be thought of in the sense of careless or poor craftsmanship. It is quite the contrary. It should be clearly understood and well known that the planning and erection of framing require extreme care and the accurate use of tools.

*Finish Carpentry.* Figure 2 shows a typical example of what is known as *finish carpentry.* In such cases, because the items will be visible after houses have been completed, appearance is equal in importance to sound construction. Thus, carpenters must take care to make joints perfectly, avoid hammer marks, keep the work smooth, and in general make sure that the completed items have a finished and beautiful appearance.

**Figure 2.** Interior-finish carpentry.

Finish-carpentry work includes the fitting and hanging of doors and windows; cutting, fitting, and nailing interior and exterior trim; making window and doorframes; fitting and nailing siding and cornices; constructing cabinets; and making all other items which will be visible after a house is ready for use.

## IMPORTANCE OF CARPENTRY

During the construction of a house, the carpenters are recognized as the key craftsmen. There is an old saying that carpenters drive the first location stakes and finally hand the front-door keys over to the owners. That saying means all that it implies. The first step in the construction of houses is the driving of location stakes. Carpenters plan for and drive those stakes. All during the construction period, carpenters generally guide all the work.

The work of many other skilled craftsmen, such as plumbers, electricians, and plasterers, is also essential to the completion of a house. However, they all look to the carpenters for information and for the good planning which is so necessary. Thus, to the carpenter falls the task of making all construction jobs progress efficiently and well.

As far as most ordinary types of houses are concerned, carpentry work is all-important. Carpenters construct the framework, which constitutes the "backbone" of a house. Unless framing is properly done, a house cannot be safe or comfortable. Then, too, the accuracy of the framing has a great bearing on all the other work involved in residential construction. The final appearance of a house (aside from design) depends almost entirely upon the skill and judgment of the carpenter.

## CITY ORDINANCES AND CODES

A study of the growth of our cities is interesting. In most cases, starting with one small house, or several, cities gradually grew until many problems had to be met and some semblance of coordinated planning carried on. Without definitely stated city building laws the construction of houses would be carried on in ways that would all too often be unsafe, unhealthy, and unattractive. Therefore, all cities create certain rules or laws, known as ordinances or codes, to govern building practices.

*Specific Purposes.* There are many important reasons why codes are necessary. Here are a few typical examples.

FIRE PROTECTION. Every year fire causes property losses which amount to millions of dollars. Many such losses can be traced to houses which were built prior to the time when city codes specified certain rules and regulations pertaining to fireproof or fire-retarding types of construction. In closely built-up cities, one house fire can cause damage to other houses.

STRUCTURAL SAFETY. Prior to the time when codes regulated structural details, too many houses were constructed without proper regard for strength. Weak foundations were built, many structural members were too small or too widely spaced, and many other unsafe construction practices were carried on. In order to prevent all such ills, city codes specify many structural requirements.

BUILDING RESTRICTIONS. City building codes also specify the types of houses which can be built in given areas. Here, the regulations are aimed at protecting all property owners. Codes protect people who are willing to build attractive houses.

*Kinds of Codes.* Every city maintains codes which apply to all planning and structural work for houses. For example, architectural, carpentry, masonry, electrical, plumbing, and other similar codes pertain to the work of all building trades. Everyone should cooperate by making sure that whatever structural work he does comes within the requirements of his local codes. Too much stress cannot be placed on the purpose and value of codes. Compliance with them means a safe and better job for homeowners.

*Inspection.* All cities employ building inspectors, whose job it is to make sure that all code provisions are complied with wherever houses are being built. Inspectors have a legal right to inspect whatever structural work is being done.

*Permits.* Most cities also require that permits be obtained before any structural work is started. Generally, working drawings, which show the extent of the work, are also required. Exact requirements, relative to permits, vary from city to city.

*Code Details.* The exact details of city codes cannot be discussed in detail since they vary considerably. Copies of codes, however, can be obtained and studied prior to the time architectural and structural work is started. The procedure is simple. Go to the local city hall or building inspector's office and read the code relative to residential construction. Detailed written information will be given. Carefully follow all the instructions received. At required periods, call in the inspector for his approval. Be careful not to cover up items which the inspector must check.

## PRESENTATION PLAN OF THIS BOOK

All the explanations in the various chapters assume that readers possess or will acquire the tool skills needed to execute all phases of carpentry work. Tool skills come only after experience and practice. Inexperienced readers can acquire such skills if they will study the directions and then practice with their tools. For example, planing skill can be acquired by using a plane on pieces of waste lumber. Skill in joinery can be acquired by actually making the joints explained in Chapter 4.

The first four chapters of the book present much information of a fundamental nature. All the explanations set forth serve to supply the reader with a basis for learning how to do good carpentry work.

Chapters 5 through 9 deal with commonly encountered carpentry work and include explanations of all important phases of house construction.

# CHAPTER TWO

# Lumber

The general subject of lumber, covering the period from the planting of the trees to the final steps in the logging and milling required to make lumber marketable, is well-nigh inexhaustible. The purpose of this chapter is to discuss the kinds of lumber, set forth the necessity for dry lumber, explain grading, caution against defects and blemishes, introduce plywood, show how board measure is calculated, and illustrate pricing. With such information, the reader will be able to select and work with wood to the best advantage.

## KINDS OF LUMBER

Lumber is classified generally into two principal kinds: softwoods and hardwoods. Oddly enough, the term *softwood,* as used in the lumber industry, does not always refer to a tree whose wood is soft. In like manner, the term *hardwood* does not always refer to a tree whose wood is hard. Actually, there is no definite or exact means of determining the difference between many trees in respect to softness or hardness. Many of the so-called softwood trees, however, have wood which is harder than that of several of the so-called hardwood trees.

In order to classify trees, in terms of soft- and hardwoods, the lumber industry has developed the custom of calling all coniferous trees softwood and all broad-leaved trees hardwood. Coniferous trees have leaves which look like needles. Such trees are popularly called evergreens. The broad-leaved trees are known as *deciduous* because they shed their leaves once a year.

**Softwoods.** The lumber industry includes the following trees in the softwood group:

| Cedars | Hemlock | Spruce |
|--------|---------|--------|
| Cypress | Larch | Tamarack |
| Douglas fir | Pines | |
| Firs | Redwood | |

**Hardwoods.** The lumber industry includes the following trees in the hardwood group:

| Ash | Butternut | Gum |
|-----|-----------|-----|
| Basswood | Cherry | Maple |
| Beech | Chestnut | Oak |
| Birch | Elm | Walnut |

**Other Trees.** Both soft- and hardwoods are obtained from many other kinds of trees. The trees listed here, however, are those whose woods are employed for most rough- and finish-carpentry work. Tables 1 and 2, at the end of this chapter, list many commonly encountered rough-, exterior-, and finish-carpentry items, together with the types of lumber generally used for them.

**Availability of Various Kinds of Lumber.** The lumberyards throughout the country generally stock the kinds of lumber which are grown and cut closest to them. This practice saves on transportation expense and makes for more economical lumber prices. There is no disadvantage in that practice because, as shown in Tables 1 and 2, the various carpentry items can be constructed from more than one kind of lumber.

## LUMBER MUST BE DRY

At the time lumber is cut from trees, it is said to be "green." In other words, it contains a considerable amount of moisture. Until much of that moisture has been disposed of, the lumber is not suitable for either rough or finish carpentry. As the moisture content diminishes, lumber changes in size. Actual shrinkage takes place, which reduces the over-all size. For example, when a green 2 by 4 dries, the shrinkage reduces both the 2-inch and the 4-inch dimensions. The exact amount of shrinkage depends upon the kind of wood and the geographical location. The length of the 2 by 4 does not change much during the drying process.

**Why Lumber Must Be Dry.** Figure 1 shows a typical example of house framing and several rough-carpentry items. If the joists, sills, plates, studs, and rafters are all green at the time of construction, some undesirable results will occur as gradual drying takes place.

First, nails do not have full holding power when driven into green wood. They are likely to withdraw partially and thus cause loose joints, squeaking, and even unsafe conditions.

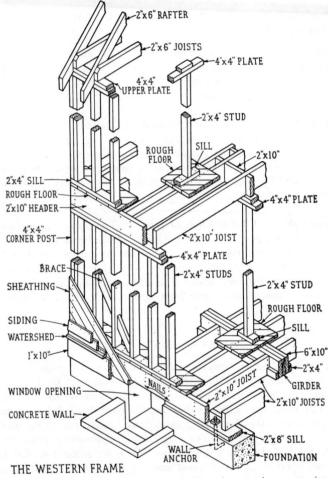

**THE WESTERN FRAME**

**Figure 1.** Typical Western framing which includes various rough-carpentry items in which shrinkage could take place.

Second, as the structural items reduce in size, the frame may warp and become out of plumb; the result is sometimes severe plaster cracking and other undesirable and unsightly conditions. As joists, sills, plates, and girders shrink in size, the items they support settle. If equal shrink-age could be depended upon, the effects would not be so bad. However, *unequal* shrinkage generally occurs, with the result that the frame may be pulled out of plumb, floors may become out of level, the trim may pull away and make cracks, doors may bind, and many other almost tragic reactions may occur.

Dry lumber is especially necessary for all finish carpentry, wherever fine cabinets, beautiful floors, or attractive joints are required. Unless all lumber is dry to start with, such items become unsightly because of open joints and poorly fitting doors and drawers.

***How Lumber Is Dried.*** There are two commonly employed methods of reducing moisture content. Either method, if properly carried out or applied, will prepare any kind of lumber for safe use.

NATURAL WAY. For structural usage, or rough carpentry, the lumber can be air-dried by placing it in piles where there are horizontal pieces between each layer. This permits a free circulation of air throughout the pile, so that drying gradually takes place.

KILN WAY. Lumber to be used in connection with finish carpentry is generally kiln-dried in order to prevent any possible shrinkage after it has been assembled into cabinets, floors, trim, etc. The kilns are large ovens in which controlled heat, over a given period of time, can be used to drive out moisture.

CAUTION. Once lumber has been dried, it should be kept in that condi-tion. Whenever it is stored out in the open, some means of protection from rain or snow should be provided. In hot climates, extreme heat from the sun can cause checking or splitting unless the lumber is stored in piles and the ends painted.

***How to Recognize Dry Lumber.*** There are several ways of recognizing dry lumber, any one or more of which can be practiced without having to resort to scientific tests. The following practical rules have been found worthy of attention.

1. Dry lumber generally weighs less than green lumber.
2. Dry lumber produces a hollow sound when rapped with the knuckles of the hand.
3. Green lumber has a sharper odor.
4. Dry lumber has more springiness.
5. Green lumber feels damp to the touch.
6. Dry lumber is likely to have small and brittle slivers on it.

## GRADES OF LUMBER

**Definitions.** Before discussing the several aspects of lumber grading, the following standard terms should be described as given and defined below.

*Boards.* Lumber which is less than 2 inches thick and at least 8 inches wide.

*Dimension lumber.* Pieces of framing lumber at least 2 inches thick and of any width.

*Dressed lumber.* Lumber which has been machine-planed. When lumber comes from the sawmill, it is rough. The next step in its preparation for use is planing to make it smooth. It may be planed (dressed) on one face or both faces, or on both faces and both edges. Several designations are used:

S1S—Lumber dressed on one face.

S2S—Lumber dressed on two faces.

S4S—Lumber dressed on both faces and both edges.

*Sizing.* When lumber is dressed or sized it is reduced from the size it was as it came from the sawmill. For example, a piece 2 inches thick and 10 inches wide, known as a 2 by 10, is reduced by planing or sawing to $1\frac{3}{4}$ by $9\frac{1}{2}$ inches. However, it is still called a 2 by 10. All dimension lumber is thus reduced somewhat in size. Such reductions must be taken into consideration when planning or doing any rough- or finish-carpentry work.

*Strips.* Pieces less than 2 inches thick and less than 8 inches wide.

*Structural timber.* Lumber which is over 5 inches in both thickness and width.

*Yard lumber.* Any lumber which is less than 5 inches thick.

**Softwood Yard Lumber.** There are several ways in which softwood yard lumber is graded.

AMERICAN STANDARD GRADES. Table 3, at the end of this chapter, shows the various grades of lumber from the standpoint of quality. The terms used in the table are standard.

STANDARD LENGTHS. Most yard lumber is sold in what are called standard lengths. All such lengths are multiples of 2. Generally, the lengths start at 6 feet and go up to 20 or more feet. In some cases, odd lengths are available. For example, studs can often be obtained 9 feet long; however, 8-foot stud lengths are standard in most lumberyards.

STANDARD WIDTHS AND THICKNESSES. Table 4, at the end of this chapter, shows typical sizes.

*Hardwood Lumber.* The best grades of hardwood lumber are known as *firsts,* the next best grades as *seconds.* Firsts and seconds, generally written FAS, practically always are combined in one grade. The third grade is termed *select,* followed by *No. 1 common, No. 2 common, sound wormy, No. 3A common,* and *No. 3B common.*

STANDARD LENGTHS. Because of their scarcity, hardwoods are standard in both odd- and even-foot lengths, such as 5, 6, 7, 8, and 9 feet.

STANDARD WIDTHS AND THICKNESSES. The widths of hardwoods have never been standardized because of the scarcity of the lumber. However, thicknesses have been standardized as shown in Table 5, at the end of this chapter.

*Miscellaneous Lumber.* Some varieties of lumber, which are used for specific purposes, have special standards which should be noted.

FINISH LUMBER. Finish lumber, used for fine cabinets, can generally be purchased S4S at most lumberyards. It must be remembered, however, that so-called 1-inch stock actually measures only $\frac{3}{4}$ inch and that $1\frac{1}{4}$-inch stock actually measures only 1 inch *net.*

The term *net* means exact measurements after dressing. Thus, if a piece of wood exactly $\frac{5}{8}$ by 7 inches is required, net dimensions must be specified.

KNOTTY PINE. Generally, knotty pine comes $\frac{3}{4}$ inch thick and $5\frac{1}{2}$, $7\frac{1}{2}$, $9\frac{1}{2}$, or $11\frac{1}{2}$ inches wide. The boards are known as 1 by 6, 1 by 8, 1 by 10, and 1 by 12.

RED CEDAR. In general, red cedar can be purchased $\frac{3}{8}$ inch thick, $3\frac{1}{4}$ inches wide, and in random lengths. It comes in bundles of about 40 board feet (see page 14). One such bundle will cover an area of about 30 square feet.

## DEFECTS AND BLEMISHES

Both defects and blemishes occur naturally in lumber. Growing trees are subject to all sorts of action by wind, rain, heat, and cold. Thus, not all lumber is perfect for rough or finish carpentry.

Defects are irregularities which reduce the strength or durability of lumber. Blemishes are marks or mars which reduce the beauty of finish lumber. In order to be sure that the lumber is good, it should be ordered and purchased according to the grades shown in Table 3. Table 6, at the end of this chapter, indicates recommended stock sizes for various purposes.

## P L Y W O O D

Plywood is a built-up board made of laminated veneers (thin layers of wood) in which the grain of each layer is at right angles to the piece next to it. The various layers are glued together under high pressure, which makes the boards as strong, or stronger, than boards composed all of one piece. The plywood boards can be purchased in various thicknesses and sizes, such as ⅜ by 30 by 60 inches or ½ by 48 by 96 inches.

Plywood can be secured in a number of different woods. *Good 2 sides* means that either surface is usable as a finish surface. *Good 1 side* is often used for drawer bottoms, where only one side has to serve as a finished surface. When so desired, plywood can be secured with a different wood on each face. This would permit matching a door panel, for instance, with the finish of two rooms, each of which is finished with a different hardwood.

Plywood has many uses and enjoys the following advantages:

1. It is comparatively light in weight.

2. Screws or nails can be driven close to the edge without danger of splitting the wood.

3. It rarely buckles or twists.

4. It does not shrink or swell.

5. It permits the construction of large flat surfaces requiring only horizontal and vertical nailing supports.

6. It can be secured in a number of different woods to match other woodwork.

7. It can be secured in white pine, which provides a fine surface for paint.

### H O W   L U M B E R   I S   S O L D

Lumber is sold by the board foot. A board foot of lumber is, basically, a piece 1 inch thick, 12 inches wide, and 12 inches long. In other words, it is 144 squares inches of lumber 1 inch thick. A piece of lumber 3 inches thick, 12 inches wide, and 12 inches long thus contains 3 board feet. In like manner, a piece 1 inch thick, 6 inches wide, and 24 inches long contains 1 board foot because its area, $6 \times 24$, is 144 square inches (see Figure 2).

The number of board feet in lumber of any dimension can be found by applying a simple mathematical formula:

$$T \times W \times L = \text{board feet}$$

where $T$ = thickness of lumber in *inches*.

$W$ = width of lumber in *feet*.

$L$ = length of lumber in *feet*.

The width of the lumber in inches is divided by 12 to obtain the width in feet.

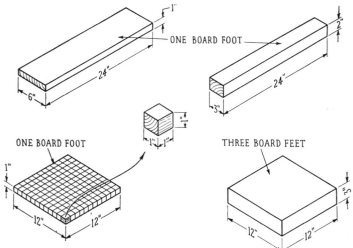

**Figure 2.** Various combinations of lumber dimensions which equal 1 board foot of lumber; 3 board feet is also illustrated.

EXAMPLE 1. Find the number of board feet in a 2 by 4 which is 12 feet long.

SOLUTION. Substitute numerical values for the letters in the formula. The value of $W$ is $\frac{4}{12}$. The value of $L$ is 12 feet. The value of $T$ is 2 inches. Thus,

$$2 \times \tfrac{4}{12} \times 12 = \text{board feet}$$

To solve the formula, first multiply 2 by $\frac{4}{12}$. That is the same as multiplying $\frac{4}{12}$ by 2. To multiply a fraction by a whole number, simply multiply the numerator of the fraction (4 in this case) by the whole number (2 in this case):

$$\tfrac{4}{12} \times 2 = \tfrac{8}{12}$$

Next, multiply $\frac{8}{12}$ by 12, in the same manner:

$$\tfrac{8}{12} \times 12 = \tfrac{96}{12}$$

To change the $\frac{96}{12}$ to a whole number, divide 96 by 12:

$$96 \div 12 = 8$$

Thus, a 2 by 4 that is 12 feet long contains 8 board feet of lumber.

EXAMPLE 2. Find the number of board feet in a plank which is 3 inches thick, 12 inches wide, and 14 feet long.

SOLUTION. Proceed as in Example 1, using the formula:

$$T \times W \times L = \text{board feet}$$

Substituting,

$$3 \times 1\tfrac{2}{12} \times 14 = 3 \times 1 \times 14 = 42 \text{ board feet}$$

In computing board feet, the dimensions of a piece of lumber are always considered on the basis of the size before dressing (surfacing). Thus, a piece $1\tfrac{5}{8}$ inches thick would be figured as 2 inches thick.

The basis for figuring plywood is much simpler, for here it is necessary to calculate only the number of square feet. This is done by multiplying the length in feet by the width in feet.


## HOW LUMBER IS PRICED

Lumber is priced on the basis of 1,000 board feet, or as so much per M. For example, one grade of lumber may be priced at $80 per M. If lumber dealers simply quote prices as so much per M, they mean so much per thousand board feet.

To determine the cost of lumber, simply multiply the number of board feet by the price per M, and divide the result by 1,000.

EXAMPLE 3. What is the cost of 32 pieces of 2 by 8 stock, each 18 feet long, if the lumber sells at $90 per M?

SOLUTION. First reduce the stock to terms of board feet. In this case, there are 24 board feet in each piece. In 32 pieces, there are 768 board feet (24 times 32).

Next, multiply the total board feet by 90:

$$768 \times 90 = 69,120$$

Finally, divide the result by 1,000. A simple way of doing this is to point off three places, starting at the right end of the figure. The 69,120 thus becomes 69.120 or $69.12, which is the cost of the lumber.

**Table 1. Kinds of wood which can be used for rough and exterior-finish carpentry**

| *Items* | *Kinds of wood* |
|---|---|
| *Rough* | |
| Framing members:<br>  Beams and girders<br>  Joists<br>  Bridging<br>  Subflooring<br>  Plates<br>  Studs<br>  Braces<br>  Ceiling joists<br>  Stiffeners<br>  Rafters<br>  Sheathing | White fir<br>Eastern hemlock<br>Western hemlock<br>Douglas fir<br>Western larch |
| Greenhouses | Western red cedar<br>Southern cypress<br>Redwood |
| *Exterior Finish* | |
| Cornice work | Redwood<br>Western red cedar<br>Southern white cedar<br>3-ply Douglas-fir paneling |
| Frames | Douglas fir<br>Ponderosa pine<br>Sugar pine<br>Yellow poplar<br>Redwood |
| Millwork:<br>  Belt course<br>  Water table | Redwood<br>Western red cedar<br>Southern white cedar |
| Porch work | Southern cypress<br>Redwood |
| Siding | Port Orford cedar<br>Western red cedar<br>Southern white cedar<br>Southern cypress<br>Western hemlock<br>Magnolia<br>Yellow poplar<br>Sitka spruce |
| Shingles | Port Orford cedar<br>Western red cedar<br>Northern white cedar |

SOURCE: Based on "Wood Handbook," Forest Products Laboratory, U.S. Government Printing Office, Washington, D.C.

**Table 2. Kinds of wood which can be used for interior-finish carpentry**

| Items | Kinds of wood |
|-------|---------------|
| Drain boards | Sugar pine |
| Drawer sides<br>Drawer bottom | Sycamore<br>Douglas fir or white pine (3-ply) |
| Flooring | Beech<br>Douglas fir (not a hardwood)<br>Western hemlock (not a hardwood)<br>Maple<br>Oak<br>Sitka spruce<br>Tupelo<br>Black gum |
| Millwork:<br>  *a.* Sash and doors | Southern cypress<br>Western larch<br>Sugar pine<br>Douglas fir (doors only)<br>Redwood (doors only)<br>Sitka spruce<br>Oak<br>Mahogany<br>Maple (doors only)<br>Birch (doors only) |
|   *b.* Cabinetwork | Alaska cedar<br>Hackberry (kitchen cabinets)<br>Douglas fir<br>White cedar<br>Walnut<br>Yellow pine<br>White pine |
|   *c.* Interior trim:<br>    Casings<br>    Base<br>    Jambs<br>    Apron<br>    Stool<br>    Stops<br>    Moldings | Oregon ash<br>Yellow birch<br>Alaska cedar<br>Port Orford cedar<br>Douglas fir<br>Red and sap gum<br>Western larch<br>Magnolia<br>Maple<br>Oak<br>Northern white pine<br>Western white pine<br>Yellow poplar<br>Redwood<br>Sitka spruce<br>Tupelo<br>Black gum<br>Black walnut |
| Stair treads | Oak<br>Vertical-grain Douglas fir |

SOURCE: Based on "Wood Handbook," Forest Products Laboratory, U.S. Government Printing Office, Washington, D.C.

**Table 3. Softwood-yard-lumber grades**

Total products of a typical log arranged in a series according to quality, as determined by appearance

1. *Select lumber* of a very good appearance and with very good finishing qualities
   - Suitable for natural finishes
     - *Grade A.* Practically free from defects.
     - *Grade B.* Allows a few small defects or blemishes.
   - Suitable for paint finishes
     - *Grade C.* Allows a limited number of small defects or blemishes that can be covered with paint.
     - *Grade D.* Allows any number of defects or blemishes that do not detract from a finished appearance when painted.

2. *Common lumber* containing defects or blemishes which detract from a finished appearance but which is still suitable for general utility and construction purposes
   - Suitable for use without waste
     - *No. 1 Common.* Sound and tight-knotted stock. May be considered watertight stock.
     - *No. 2 Common.* Allows large and coarse defects. May be considered grain-tight lumber.
   - Lumber permitting waste
     - *No. 3 Common.* Allows larger and coarser defects than No. 2 and some knotholes.
     - *No. 4 Common.* Low-quality lumber admitting the coarsest defects, such as rot and holes.

Table 4. Standard thicknesses and widths for softwood yard lumber

| Product | Rough-green sizes, inches | | Minimum rough-dry dimensions, inches | | Dressed dimensions, inches | |
|---|---|---|---|---|---|---|
| | Thickness | Width | Thickness, standard yard | Width | Thickness, standard yard | Width (face when worked) |
| Finish | ... | 3 | ..... | $2\frac{3}{4}$ | $\frac{5}{16}$ | $2\frac{5}{8}$ |
| | ... | 4 | ..... | $3\frac{5}{8}$ | $\frac{7}{16}$ | $5\frac{1}{2}$ |
| | ... | 5 | ..... | $4\frac{5}{8}$ | $\frac{9}{16}$ | $4\frac{1}{2}$ |
| | ... | 6 | ..... | $5\frac{5}{8}$ | $\frac{11}{16}$ | $5\frac{1}{2}$ |
| | 1 | 7 | $2\frac{9}{32}$ | $6\frac{5}{8}$ | $2\frac{5}{32}$ | $6\frac{1}{2}$ |
| | $1\frac{1}{4}$ | 8 | $1\frac{3}{16}$ | $7\frac{3}{8}$ | $1\frac{1}{16}$ | $7\frac{1}{4}$ |
| | $1\frac{1}{2}$ | 9 | $1\frac{7}{16}$ | $8\frac{3}{8}$ | $1\frac{5}{16}$ | $8\frac{1}{4}$ |
| | $1\frac{3}{4}$ | 10 | $1\frac{9}{16}$ | $9\frac{3}{8}$ | $1\frac{7}{16}$ | $9\frac{1}{4}$ |
| | 2 | 11 | $1\frac{6}{8}$ | $10\frac{3}{8}$ | $1\frac{5}{8}$ | $10\frac{1}{4}$ |
| | $2\frac{1}{2}$ | 12 | $2\frac{1}{4}$ | $11\frac{3}{8}$ | $2\frac{1}{8}$ | $11\frac{1}{4}$ |
| | 3 | | $2\frac{6}{8}$ | .... | $2\frac{5}{8}$ | .... |
| Common boards and strips | 1 | 3 | $2\frac{9}{32}$ | $2\frac{3}{4}$ | $2\frac{5}{32}$ | $2\frac{5}{8}$ |
| | $1\frac{1}{4}$ | 4 | $1\frac{3}{16}$ | $3\frac{3}{4}$ | $1\frac{1}{16}$ | $3\frac{5}{8}$ |
| | $1\frac{1}{2}$ | 5 | $1\frac{7}{16}$ | $4\frac{3}{4}$ | $1\frac{5}{16}$ | $4\frac{5}{8}$ |
| | ... | 6 | ..... | $5\frac{3}{4}$ | ..... | $5\frac{5}{8}$ |
| | ... | 7 | ..... | $6\frac{3}{4}$ | ..... | $6\frac{5}{8}$ |
| | ... | 8 | ..... | $7\frac{5}{8}$ | ..... | $7\frac{1}{2}$ |
| | ... | 9 | ..... | $8\frac{5}{8}$ | ..... | $8\frac{1}{2}$ |
| | ... | 10 | ..... | $9\frac{5}{8}$ | ..... | $9\frac{1}{2}$ |
| | ... | 11 | ..... | $10\frac{5}{8}$ | ..... | $10\frac{1}{2}$ |
| | ... | 12 | ..... | $11\frac{5}{8}$ | ..... | $11\frac{1}{2}$ |
| Dimension and heavy joist | 2 | 2 | $1\frac{3}{4}$ | $1\frac{3}{4}$ | $1\frac{5}{8}$ | $1\frac{5}{8}$ |
| | $2\frac{1}{2}$ | 4 | $2\frac{1}{4}$ | $3\frac{3}{4}$ | $2\frac{1}{8}$ | $3\frac{5}{8}$ |
| | 3 | 6 | $2\frac{3}{4}$ | $5\frac{3}{4}$ | $2\frac{5}{8}$ | $5\frac{5}{8}$ |
| | 4 | 8 | $3\frac{3}{4}$ | $7\frac{5}{8}$ | $3\frac{5}{8}$ | $7\frac{1}{2}$ |
| | ... | 10 | ..... | $9\frac{5}{8}$ | ..... | $9\frac{1}{2}$ |
| | ... | 12 | ..... | $11\frac{5}{8}$ | ..... | $11\frac{1}{2}$ |

Table 5. Standard thicknesses for hardwoods

| Rough, inches | Surfaced 1 side (S1S), inches | Surfaced 2 sides (S2S), inches | Rough, inches | Surfaced 1 side (S1S), inches | Surfaced 2 sides (S2S), inches |
|---|---|---|---|---|---|
| $\frac{3}{8}$ | $\frac{1}{4}$ | $\frac{3}{16}$ | $2\frac{1}{2}$ | $2\frac{5}{16}$ | $2\frac{1}{4}$ |
| $\frac{1}{2}$ | $\frac{3}{8}$ | $\frac{5}{16}$ | 3 | $2\frac{13}{16}$ | $2\frac{3}{4}$ |
| $\frac{5}{8}$ | $\frac{1}{2}$ | $\frac{7}{16}$ | $3\frac{1}{2}$ | $3\frac{5}{16}$ | $3\frac{1}{4}$ |
| $\frac{3}{4}$ | $\frac{5}{8}$ | $\frac{9}{16}$ | 4 | $3\frac{13}{16}$ | $3\frac{3}{4}$ |
| 1 | $\frac{7}{8}$ | $1\frac{3}{16}$ | $4\frac{1}{2}$ | (*) | (*) |
| $1\frac{1}{4}$ | $1\frac{1}{8}$ | $1\frac{1}{16}$ | 5 | (*) | (*) |
| $1\frac{1}{2}$ | $1\frac{3}{8}$ | $1\frac{5}{16}$ | $5\frac{1}{2}$ | (*) | (*) |
| 2 | $1\frac{13}{16}$ | $1\frac{3}{4}$ | 6 | (*) | (*) |

* Finished size not specified in rules. Stock usually made in small quantities or on special order.

**Table 6. Stock sizes generally used for various rough and finish items ***

*Structural items*

| *Items* | *Stock sizes* | *Items* | *Stock sizes* |
|---|---|---|---|
| Form unit: | | Wall unit (cont'd): | |
| Forms | 1 x 6 | Sheathing | 1 x 6 S1E; 1 x 6 T and G |
| Braces | 1 x 4 | Plaster grounds | ¾ x ¾ |
| Stakes | 1 x 4; 2 x 3 | Second-floor unit: | |
| Whalers | 2 x 4 | Joists and headers | 2 x 8; 2 x 10 |
| Underpinning unit: | | Long spans | 2 x 12–2 x 16 |
| Sills | 2 x 6; 2 x 8 | Bridging | 2 x 3 |
| Pier blocks | 2 x 6 x 6; 2 x 8 x 8 | Subflooring | 1 x 6 |
| Girders | 4 x 4; 4 x 6 | Ceiling unit: | |
| Beams | 4 x 10; 4 x 12 | Joists | 2 x 4; 2 x 6 |
| Posts | 4 x 4 | Stiffeners (strong back) | 2 x 8 |
| Lally columns | 3″ diameter metal | Backing | 2 x 4 |
| Cribbing plates | 2 x 4; 2 x 6 | Roof unit: | |
| Cribbing studs | 2 x 4; 2 x 6 | Rafters | 2 x 4; 2 x 6 |
| First-floor unit: | | Hips and valleys | 2 x 6; 2 x 8 |
| Joists and headers | 2 x 6; 2 x 8 | Purlins | 2 x 4 |
| If basement | 2 x 10 | Braces | 2 x 4 |
| Bridging | 2 x 3 | Gable studs | 2 x 4 |
| Subflooring | 1 x 6 | Sheathing: | |
| Wall unit: | | Solid | 1 x 6 |
| Plates and studs | 2 x 4; 2 x 6 | Spaced | 1 x 4 |
| Braces | 2 x 4; 2 x 6 | Stair unit: | |
| Cut in | 1 x 6 | Stair horse | 2 x 10 |
| Backing | 2 x 4; 1 x 8 | Rough treads | 1 x 10 |
| Fire stops | 2 x 4; 2 x 6 | | |

*Finish items*

*Exterior-trim unit*

| | | | |
|---|---|---|---|
| Belt course | 1 x 6 | Frames unit (cont'd): | |
| Water table | 2 x 3 | Parting bead | ⅜ x ¾ |
| Siding | ⅝ x 4; 1 x 6; 1 x 8; 1 x 10 | Sills | 2 x 8 |
| Cornice unit: | | Casings | 1 x 6 or 1 x 1½ plaster molding |
| Frieze | 1 x 6 | Doorjambs | 1 x 6; 1¼ x 6 |
| Fascia | 1 x 4 | Stops | ½ x 4 or rabbeted |
| Plancier | 1 x 10; 1 x 12; or T-and-G stock | Porch unit: | |
| Moldings | 1 x 2; 1 x 4 | Posts | 4 x 4; 10″ diameter |
| Frames unit: | | Railing | 2 x 10 |
| Pulley stile | 1 x 5 | Balusters | 2 x 2 |
| Blind stop | 1 x 2 | Soffit | 1 x 6 |
| | | Trim | 1 x 6 |

*Interior-trim unit*

| | | | |
|---|---|---|---|
| Porch unit: | | Cabinet unit: | |
| Casings and apron | ⅝ x 2; 1 x 3 | Jambs and shelves | 1″ stock |
| Baseboard | 1 x 2; 1 x 4; 1 x 6 | Drawer sides | ⅝″ stock |
| Doorjambs | 1 x 6; 1¼ x 6 | Face frame | 1″ net frames |
| Stool | 1¼ x 4 | Stair unit: | |
| Doorstops | ⅜ x 1½ | Treads | 1″ net thickness |
| Floor unit: | | Stringers | 1″ stock |
| Hardwood | ½ x 1½; ½ x 2; ¾ x 3 | Handrail | 2 x 3; 3 x 3 |
| Plywood (under linoleum) | 48 x 96 x ⅜ | Balusters | 1 x 1 net to 1¾ x 1¾ |

* The above is not a complete list. For the sake of economy, the sizes of many parts are based, whenever possible, on stock sizes of lumber. The span of the floor joists also affects the sizes of joist stock. Pitch of roof and snow load affect sizes of roof members.

# CHAPTER THREE

# Hand Tools and Power Machinery

The hand tools used by the carpenter can be classified into two groups: in the first group, all the required power is personally supplied by the person using the tools; in the second, the power is supplied by electricity, and the operator simply guides the tools. The two groups are known, respectively, as *hand tools* and *powered hand tools*.

*Power machinery* is the term applied to machines which must be fastened down, as to a floor, and which derive all their power from electricity. Such machines are increasing in popularity and are used for all sorts of carpentry and cabinetwork.

It should again be emphasized that the good carpenter is also a good cabinetmaker—hence the emphasis placed in this book on fine precision work, even though, in trade practice, some of the operations described may be considered by some as strictly cabinetmaking. Much of the work done by a cabinetmaker can also be done by a skilled carpenter.

## HAND TOOLS

For purposes of easy identification, hand tools can be named on the basis of the type of work done with them. In this section we shall discuss (1) measuring tools, (2) layout and marking tools, (3) holding tools, (4) cutting tools, (5) boring tools, (6) smoothing tools, (7) finishing tools, (8) assembling tools, (9) leveling and plumbing tools, (10) prying tools, and (11) sharpening tools.

*Measuring Tools.* There are several commonly used measuring tools, as shown in Figure 1.

POCKET RULES. Pocket rules are made of either wood or metal and are used for any kind of measuring on projects made on a bench or as part of a house. These rules are often referred to as folding or zigzag rules. The flexible push-pull 6- or 8-foot tape is also classified as a pocket rule.

STEEL TAPE. A steel tape is required for house-construction layout where measurements longer than 6 or 8 feet must be made. The 50-foot tape is long enough for most carpentry needs. A cloth tape is not considered satisfactory for accurate measuring since it varies in length according to the amount of stretching done by the user.

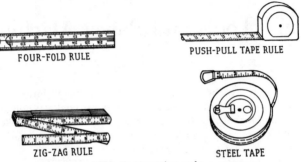

FOUR-FOLD RULE      PUSH-PULL TAPE RULE

ZIG-ZAG RULE      STEEL TAPE

**Figure 1.** Measuring tools.

*Layout and Marking Tools.* Basic tools used primarily for layout work include the try square, combination square, pocketknife, steel square, straightedge, dividers, scriber, T bevel, plumb bob, marking gauge, and butt gauge. These are illustrated in Figure 2.

6-INCH TRY SQUARE AND COMBINATION SQUARE. These tools are standard in most tool kits; if one has the combination square, however, the try square need not be purchased, for a combination square permits squaring, miter-cut layout, depth measuring, and parallel marking.

EXAMPLE. If a board 6 inches wide needs to be marked for ripping (by hand) to make it 4¾ inches wide, the blade of the combination square is set at this measurement, the head of the square held tightly against the edge of the board (which must be smooth and straight), a pencil held against the movable blade, and then the square moved from one end of the board to the other.

POCKETKNIFE. This tool, rather than a pencil, is always used for precision layout work, for a knife mark is always much sharper than any that can be obtained with even a hard pencil.

STEEL SQUARE. The steel square is considered one of the most adaptable layout tools, permitting the laying out of angles and cuts for roofs, stairs, braces, and any angle cut. The markings on steel squares vary; the one best suited for the woodworker is marked in twelfths of an inch, as well as in sixteenths and eighths. The twelfth markings permit mathematical layout calculations on a simple scale basis: 1 inch equals 1 foot; $\frac{1}{12}$ inch equals 1 inch.

STRAIGHTEDGE. The straightedge is a "must" for the layout man. Such a tool is made in the shop, not purchased, for the size of a straightedge depends entirely on the job to be layed out. For cabinetmaking the

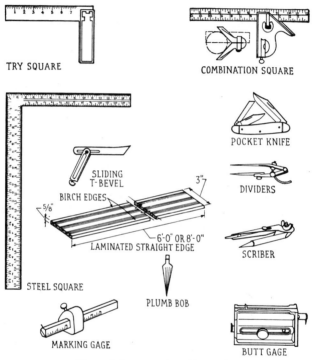

**Figure 2.** Layout and marking tools.

straightedge may need to be only 6 or 8 feet long, with thickness and width dimensions in proportion; for construction work a straightedge is often 10 or 12 feet long. Special jobs may require even longer lengths.

The best straightedge is a built-up, or laminated, one, that is, one made by gluing narrow strips of hardwood edge to edge until the required width is secured. Maple or birch is best for the outside edges. A laminated straightedge will always hold a true shape; hence, if carefully hung up when not in use (a hole should be bored in one end for this purpose,) it will always be ready for immediate use.

The purpose of a straightedge is to provide a means of marking material that is wider or longer than a steel square, the limitation of which is 24 inches. It also permits making layout lines on a large, flat surface, such as a floor, wall, or panel.

DIVIDERS. A pair of dividers is very useful for problems of spacing. Moreover, if a pencil is inserted in place of one of the metal legs, this tool can also be used as a compass in making circular layouts.

SCRIBER. A scriber is probably the least costly and yet one of the most useful tools in a carpenter's toolbox. Though shaped very much like a pair of dividers, it is used, not as a compass, but to make lines parallel to a surface against which a piece of lumber is to be fitted. The smallness of the tool permits it to follow any irregularities in the surface; hence the cutting line made is also irregular; if the lumber is cut exactly as marked, a perfect joint will result.

T BEVEL. The T bevel, or bevel square, has an adjustable blade which permits the setting of the tool to any angle required. Once the tool is set, duplicate angle cuts can be made, or if power machinery is to be used, the saw guides on the machine can be adjusted to the exact angle by means of the T bevel. The tool has blades from 6 to 8 inches in length; a 6-inch blade is suitable for most work.

The T bevel is particularly useful in rafter cutting. Once a rafter pattern has been made, the bevel is set to the exact angle and is then used to mark duplicate angles on all other rafters required.

The same principle is applied in brace cutting: the correct angle is established (sometimes by actually placing the brace stock in position and marking the exact angle and cutting line), and the bevel is then set to this angle.

PLUMB BOB. A plumb bob is a very useful layout tool, particularly when it is desired to locate a point either directly above or directly below a given point. If the tool is used out of doors, such as on a construction job, letting the plumb bob rest in a bucket of water will stop the swaying motion occasioned by a breeze or wind. In this case the layout problem is assumed to be one of establishing a vertical line from which accurate measurements can be taken.

Plumb bobs come in various sizes or weights; a "rough" one may be of cast iron; a precision tool is made from turned brass. The degree of accuracy also depends on the size of the line used.

MARKING GAUGE. A marking gauge is very useful in laying out lines parallel to the sides of the stock or locating a center line for dowels. It is a required tool when "dapping" in for hinges, for it enables the user to mark a very accurate and sharp line, which then serves as a guideline to follow in chiseling out the wood to receive the hinge leaf.

BUTT GAUGE. A butt gauge is used exclusively for the layout work connected with "dapping" or "gaining" in hinge seats. This tool is particularly useful when hanging a door to a rabbeted jamb, as it will automatically provide for the very necessary clearance between the face of the door and the edge of the doorstop.

*Holding Tools.* There are numerous carpentry operations which require either a holding tool to hold the cutting tool or a device to hold the job itself—or part of it. The former classification includes the hand brace and hand drill; the latter classification refers to the bench vise. All these holding tools are illustrated in Figure 3.

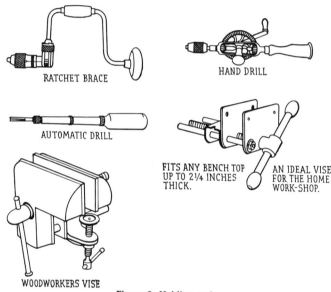

RATCHET BRACE

HAND DRILL

AUTOMATIC DRILL

FITS ANY BENCH TOP UP TO 2¼ INCHES THICK.

AN IDEAL VISE FOR THE HOME WORK-SHOP.

WOODWORKERS VISE

Figure 3. Holding tools.

HAND BRACES AND HAND DRILLS. Hand braces and hand drills are designed, respectively, to hold wood bits and drills. Braces are purchased on the basis of the "sweep" of the handle—10-inch, 12-inch, etc. The best hand brace has a ratchet device which permits a backward movement of the brace handle and thus makes it possible to bore holes in an inside angle. The hand drill, sometimes called the egg-beater type, is made in large and small sizes. The automatic hand drill has a powerful spring which permits a continuous "push action." Either tool is good, but the automatic drill is easier to use in tight spots; if it is not used carefully, however, the spring action may cause the drill to jump out of the hole and mar the wood.

BENCH VISES. Bench vises are of two types, as illustrated in Figure 3: (1) an all-metal vise with an adjustable thumbscrew, permitting it to be fastened to the end of the bench top, the end of a sawhorse, the end of a shelf, etc.; (2) a vise with single or double screw shafts that is built into a bench top. This vise has hardwood (birch or maple) faced jaws and is most useful to the carpenter or cabinetmaker who works continuously at a workbench. The metal vise is best adapted to small objects, the bench vise to long pieces of lumber which need to be jointed.

*Cutting Tools.* (See Figure 4.) Cutting tools are varied in character and purpose. They include saws, chisels, and gouges. The term *cutting*, as used here, refers not only to making two or more pieces from one piece, by means of a saw, but also to dapping hinges (cutting in a flat recess exactly to receive one leaf of a hinge) and shaping various parts of a project by means of a chisel or gouge.

SAWS. Handsaws, in general, can be classified as either crosscut, for cutting across the grain of the wood; or ripsaws, for cutting with the grain of the wood.

The basic difference between the two types is in the shape of the teeth, as indicated in Figure 4. Handsaws are made with either a straight or curved (skew) back. Selection is a matter of personal preference.

Each type of saw blade is filed to fit the purpose for which it is made. Saws are designated by the number of teeth per inch, as 7-point, 8-point, etc. The higher the number, the finer the teeth; the finer teeth are used in finish carpentry.

There are several other kinds of saws made to do special work. These are the backsaw, coping saw, keyhole saw, miter-box saw, and hack saw. Each is described below.

*Crosscut saws.* The carpenter should have two crosscut saws: one 8-point saw for cutting rough or green lumber and one 10- or 11-point saw for finish work. Each should be 26 inches in length.

*Ripsaws.* Ripsaws are made 24 or 26 inches in length. It is possible to buy a 5½-point ripsaw, but the more common 7- or 8-point saw is satisfactory for general work.

*12-inch backsaw.* The 12-inch backsaw, shown in Figure 4, is very useful for fine cabinetwork. It has very small teeth designed for fine cutting. The number of teeth per inch is established by the manufacturer.

*Coping saw.* The coping saw, illustrated in Figure 4, is required for light scrollwork or for coping the ends of moldings to make an inside-angle joint.

*Keyhole saw.* Usually purchased with detachable blades, of three different lengths, which can easily be removed and replaced on the handle, the keyhole saw is useful for cutting inside circles or small curved corners or angles.

*Miter box and saw.* A miter box, actually a metal frame carefully designed to hold vertical posts in which the saw slides, permits fine dupli-

COPING SAW

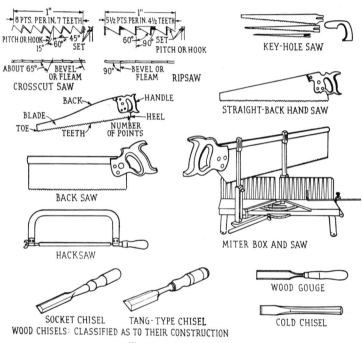

**Figure 4.** Cutting tools.

cate precision cutting for any angle desired on material up to the width capacity of the box, as illustrated in Figure 4. This tool is not a necessity for the average mechanic; in construction work, the contractor often furnishes a miter box for his workmen to use, because of its high cost and the fact that it is too large to be carried in the average carpenter's hand or toolbox.

*Hack saw.* A hack saw is a "must" for the mechanic who does a great variety of work. This tool is used for cutting metal, such as bolts and nails, and is constructed so that the blade can be replaced as it becomes worn. A screw device at one end permits giving the saw blade the proper tension.

**Note:** During the framing processes it sometimes becomes necessary to move a framed member after it has been securely nailed. The hack saw, because of its thickness, cannot be used. A very excellent substitute can be made by simply filing a set of fine teeth on about 3 inches of the lower end of the back edge of the regular crosscut handsaw. The teeth are filed with a three-cornered file at right angles to the face of the blade and the back edge of the blade. A dozen nails can be cut before refiling is necessary. This "nail saw" has many advantages, for it is always ready for immediate use.

CHISELS. There are two general types of chisels, framing and finish. The handle of a chisel is fitted to the metal shank by one of two methods: socket or tang. In the socket type, the handle fits *into* a socket which is actually part of the blade. In the tang type, the tang of the chisel fits into the handle. The socket type will withstand pounding better than the tang type (see Figure 4).

Chisel blades vary in width from ⅛ inch to 1 inch (by eighths) and from 1 inch to 2½ inches (by quarters). The length of the blade and shank will vary according to the width of the chisel and the purpose for which the chisel was designed. Hence it is possible to secure a long framing chisel (socket type) only ⅛ inch in width and a quite short finish chisel that is 2 inches wide.

Chisels can be secured in sets in attractive wooden boxes. This is ordinarily the most satisfactory way to buy chisels; it is more useful, however, to have a variety of chisels designed for specific purposes than to own a complete set of one type. It is possible, of course, to acquire complete sets of several different types, but this practice is not usually followed.

Chisels can be classified, as noted above, into framing and finish types. The latter category includes paring, butt, pocket, and mortise chisels.

*Framing chisel.* Framing chisels, which are of the socket type, are constructed with long, extra-heavy shanks to permit rough cutting during the erection of a framed building.

*Paring chisel.* The paring chisel has a short blade to permit making paring cuts. A paring cut is made by holding the back of the chisel perfectly flat to the wood surface and then moving the chisel sideways and pushing it gently forward. This cutting action results in a thin shaving and gives a smooth, flat surface.

*Butt chisel.* The butt chisel is made 2½ inches wide and very short—approximately 3 inches. This tool is particularly useful when cutting (dapping in) hinge seats to receive the leaf of a hinge during the process of hanging a door or casement sash (see Figure 4). The short, broad blade makes it quite easy to take paring cuttings, thus assuring a perfectly flat surface on which the leaf of the hinge will lie.

*Pocket chisel.* A pocket chisel has a blade about 4½ inches long and is useful when cutting hardwoods.

*Mortise chisel.* The mortise chisel is a long, narrow tool with a thick blade, which makes it very adaptable for cutting mortises by hand.

GOUGES AND CARVING TOOLS. Gouges and carving tools are made in various sizes, with inside or outside bevels, and with straight shanks or bent shanks. Carving tools also come in a variety of shapes and sizes. These tools are not a necessity for the average worker in wood, but anyone who desires to do fine carving and shaping should study these tools at the hardware store and then make his selection.

COLD CHISEL. A cold chisel, a very necessary tool, is used for cutting cold steel—hence its name. Cold chisels are made in various lengths, from 5 to 18 inches, and with cutting edges of different widths, from $\frac{5}{16}$ to 1¼ inches. For light work a 6- by ½-inch chisel is recommended; for heavy work the chisel should be 8 inches long and have a 1-inch cutting edge.

**Boring Tools.** (See Figure 5.) Boring tools are classified according to the nature of the work that is done with them. Typical boring tools that are used to make holes are augur bits, drill bits, expansive bits, countersinks, and star drills.

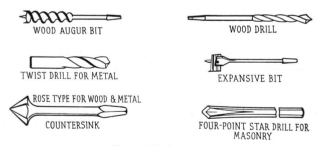

WOOD AUGUR BIT

WOOD DRILL

TWIST DRILL FOR METAL

EXPANSIVE BIT

ROSE TYPE FOR WOOD & METAL
COUNTERSINK

FOUR-POINT STAR DRILL FOR MASONRY

**Figure 5.** Boring tools.

AUGUR BITS. For boring holes in wood, augur bits are available that vary in size from $\frac{3}{16}$ to 1⅛ inches (by sixteenths) and from 1⅛ to 2 inches (by eighths). The number stamped on the shank indicates its size. A No. 10 bit will make a ⅝-inch hole; a No. 6 bit, a ⅜-inch hole; etc. The feed screw at the end of the bit determines the speed of the bit; a coarse thread is known as a fast bit; a fine thread will cut much slower. For precision work the slow bit is recommended.

WOOD DRILL. As its name implies, the wood drill is used for wood only. The skilled carpenter will usually have several of the smaller sizes, such as ⅛-inch, ¼-inch, etc.

TWIST DRILL. The twist drill is designed for drilling in metal only. It can be used in shallow-hole drilling, and in electric, hand, and breast drills. Obviously, if the twist drill is, say, ¾ inch in size, it must be used with a power drill in order to bore a hole in metal.

EXPANSIVE BIT. The expansive bit is so called because it can be expanded to cut holes of various sizes (in wood only). This bit is particularly useful in fitting cylinder locks or for boring holes to receive metal pipes, etc. It can be purchased with two detachable expansive bits, each of a different size, permitting the cutting of holes up to 3 inches in diameter.

COUNTERSINK. A countersink is a necessity for any woodworker, as it permits the cutting of a slanting recessed hole to receive the head of a flat-headed screw and make it flush with the surface of the wood.

STAR DRILL. The star drill is a very useful tool for making holes in masonry to receive a bolt sleeve or a metal sleeve. Star drills come in several different sizes, which are designated by the diameter measurements as ¼-inch, ⅜-inch, etc.

*Smoothing Tools.* (See Figure 6.) Wood-smoothing tools are known as planes and can be purchased in a large variety of sizes and types. Each type has its own special use or purpose. The discussion below is limited to those planes which are considered an absolute necessity for the beginner. Skilled carpenters and cabinetmakers will have several others in order to simplify and speed up their daily work.

Each workshop toolbox should contain a jack plane, smoothing plane, fore plane, block plane, and spokeshave, as illustrated in Figure 6. The other planes described below are not absolutely necessary.

JACK PLANE. The jack plane is an all-purpose plane, although it is generally used to remove the surface of a rough piece of lumber, after which the smoothing plane is used to give a very smooth surface.

SMOOTHING PLANE. The smoothing plane is used for the final smoothing of wood prior to sanding. The setting of the breaker on the plane bit determines the fineness of the shaving. If this tool is kept in sharp condition, much time and effort are saved in sanding.

FORE PLANE. The fore plane which is 18 inches in length serves admirably as a jointer on short stock. It should not be used for hanging doors, except small cabinet doors.

JOINTER. A jointer is made either 22 or 24 inches in length and, as the name implies, is designed for making the edges of boards straight prior to gluing them together. The tool is also a necessity for full-size-

JACK PLANE

SMOOTHING PLANE

JOINTER

FORE ·PLANE

ROUTER PLANE

BLOCK PLANE

SPOKE SHAVE

**Figure 6.** Smoothing tools.

door hanging. Since a jointer weighs several pounds, the construction carpenter will usually not have one unless he does a lot of door hanging or makes a lot of built-in cabinets.

BLOCK PLANE. The block plane is a small tool designed for planing end-grained wood or very small pieces of wood. It is approximately 6 inches in length and $1\frac{1}{2}$ inches in width. The larger planes have a plane bit or plane iron and plane cap; a block plane has no plane cap.

SPOKESHAVE. Occasionally a curved surface will need smoothing which cannot be done with an ordinary plane. The spokeshave (its name indicates its original purpose) will enable the carpenter to do a first-class job of cleaning such a surface. The straight-bottomed spokeshave is probably the best tool for general purposes; however, spokeshaves with convex or concave bottoms can be secured.

ROUTER PLANE. One more plane should be mentioned, even though it is not found in many tool kits. The router plane is used when leveling the bottom of a dado or groove made to receive the end of a shelf. Since this tool greatly simplifies the making of certain types of precision joints, there is a great advantage in adding it to a tool kit.

*Finishing Tools.* After a piece of finish stock has been planed to remove imperfections, it is often necessary to do more "finishing" to prepare the stock for the stain, varnish, lacquer, or other materials used to preserve the surface of the wood and give it an attractive appearance.

CABINET SCRAPER. The cabinet scraper is fitted with a removable beveled-edge scraper blade, which—when properly sharpened by turning the sharp corner of the beveled edge with a burnisher (see Figure 11)—will smooth ridges or torn grain left by the smoothing plane.

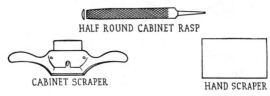

HALF ROUND CABINET RASP

CABINET SCRAPER

HAND SCRAPER

**Figure 7.** Finishing tools.

HANDSCRAPER. The handscraper is a piece of steel 2 or 3 inches in width and 4 to 6 inches in length. The sharp corners of this tool are sharpened by the use of a burnisher and will actually remove a very fine shaving. It is especially adaptable in close quarters, where it would not be possible to use the cabinet scraper.

HALF-ROUND CABINET RASP. The half-round cabinet rasp, which can be secured in various grades of teeth (fine to coarse), is useful in smoothing curved edges or corners. For finish work, only the fine-tooth rasp is necessary.

These finishing tools are relatively inexpensive and should be a part of any tool kit if fine cabinetwork is to be done.

*Assembling Tools.* Tools used for assembling a project include screwdrivers, hammers, hand ax, mallets, wrenches, pliers, nail sets, C clamps, hand screws, bar clamps, level and plumb, and pinch bar, as illustrated in Figure 8.

SCREWDRIVERS. Screwdrivers vary in length and in width of tip. Ratchet-type screwdrivers permit a return action of the hand without the user having to remove his hand from the handle. Still another type is the spiral-ratchet screwdriver. It *is* also possible to secure screwdriver bits

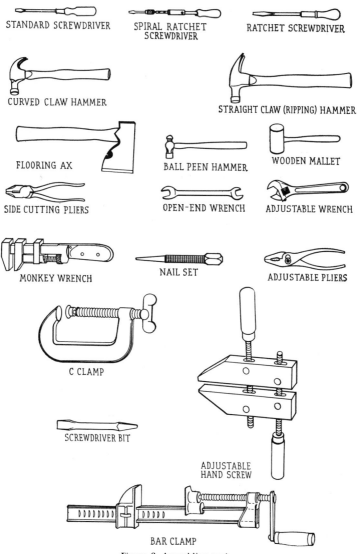

STANDARD SCREWDRIVER

SPIRAL RATCHET
SCREWDRIVER

RATCHET SCREWDRIVER

CURVED CLAW HAMMER

STRAIGHT CLAW (RIPPING) HAMMER

FLOORING AX

BALL PEEN HAMMER

WOODEN MALLET

SIDE CUTTING PLIERS

OPEN-END WRENCH

ADJUSTABLE WRENCH

MONKEY WRENCH

NAIL SET

ADJUSTABLE PLIERS

C CLAMP

SCREWDRIVER BIT

ADJUSTABLE
HAND SCREW

BAR CLAMP

**Figure 8.** Assembling tools.

which will fit into the bit brace. This particular type is very useful when turning large-diameter screws, for additional leverage is secured through the use of the ratchet brace.

The average tool kit should include several standard screwdrivers of varying lengths, such as a 4-inch, 6-inch, and 8-inch; one ratchet-spiral screwdriver, if a great deal of screw work is to be done, and by all means, at least two screwdriver bits of different sizes.

HAMMERS. Hammers are classified as straight-claw (very useful in rough carpentry and form building), curved-claw, and ball-peen. The latter is used primarily for metalwork.

Claw hammers vary in size, which is designated by weight in ounces. The 16-ounce hammer is standard for most framing jobs, and the 12-ounce is used on fine cabinetwork.

HAND AX. A hand ax is a necessity, if concrete forms are to be constructed, in order to sharpen the stakes. It is also used in rough-carpentry work to chop off surplus material, such as the end of a ceiling joist which would otherwise protrude through the roof.

WOODEN MALLET. When a considerable amount of chiseling is to be done, a wooden mallet should be used to save the end of the chisel handle. Considered a finish tool, it would not be used in heavy construction work.

WRENCHES. Wrenches are classified as open-end adjustable, monkey, and S-type. Each end of the S-type wrench is made to fit a nut of a particular size; hence S wrenches are often bought in sets. The size of a monkey wrench or an open-ènd wrench is designated by its length, which may vary from 4 to 20 inches. The 8-inch wrench would probably be the most useful size for an average workshop.

PLIERS. Pliers are made in many shapes and sizes and for many particular purposes. The two pairs of pliers illustrated in Figure 8 are typical examples of the types most adaptable for the craftsman. The side-cutting pliers provide a very excellent tool for cutting wire. The combination plier is useful for tightening nuts on bolts when a wrench is not practical.

NAIL SETS. Nail sets are designed for setting nailheads below the surface of the wood to permit them to be covered with putty by the decorator. They are procurable in various sizes; the term *size*, as used here, indicates the size of the end placed on the nailhead. The selection of the desired size is therefore based on the size of the nailheads to be set. One small-sized and one medium-sized nail set are sufficient for the average tool kit.

C CLAMPS, HAND SCREWS, AND BAR CLAMPS. The C clamp and hand screw are useful for holding or gluing small pieces of lumber together to make a piece of the required thickness. The bar clamp is most adaptable for gluing pieces edge to edge to get greater width. A complete kit of tools should include at least two 6-inch C clamps and two 4-foot bar clamps (dimensions indicate capacity).

*Leveling and Plumbing Tools.* A most important tool in the carpenter's kit is the level and plumb, which is actually one tool with level glasses so placed that they enable the mechanic to test a job for either levelness or plumbness. It is by levelness and plumbness, as well as squareness, that the work of a craftsman is most easily judged, even by a novice. Hence the level and plumb should be selected most carefully. Some level glasses are replaceable; this is a most desirable feature, as it assures long life of the tool.

The tool illustrated in Figure 9 is called a level, but it will also serve as a plumb. Note the single leveling glass and the two plumbing glasses.

ALUMINUM ADJUSTABLE LEVEL

**Figure 9.** Leveling and plumbing tool.

This tool, securable in wood, cast iron, and aluminum, comes in varying lengths. The brickmason's level is 4 feet in length. The carpenter's level is 24 or 30 inches in length, which is long enough for any carpentry job, since when necessary, the level or plumb will be used in connection with a straightedge. It is recommended that the carpenter have a 24-inch level, which can be either metal or wood.

*Prying Tool.* An 18- or 24-inch pinch bar, as illustrated in Figure 10, is very useful for lifting the corner of a box, opening a crate, holding the

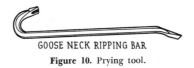

GOOSE NECK RIPPING BAR

**Figure 10.** Prying tool.

end of a piece of lumber in place, and for many other kindred operations. A nail claw on the rounded end of the bar permits the withdrawal of nails too large to pull with a hammer.

*Sharpening Tools.* (See Figure 11.) Sharpening tools divide into several groups: (1) saw clamps, (2) saw sets, (3) files for filing handsaws, (4) a special file for sharpening augur bits, (5) oilstones for sharpening plane bits and chisels, (6) a slip stone for sharpening gouges, and (7) burnishers for "turning an edge" on scraper blades.

SAW CLAMPS. Saw clamps are manufactured in various types, any one of which is satisfactory for the average user.

SAW SET. A saw set is needed to set the teeth of a handsaw to provide cutting clearance so that the saw can slide freely; rough lumber requires a saw with considerable set; finish saws have but little set. The amount of set in a saw, providing that the saw is properly filed, determines whether the resulting cuts are smooth or rough.

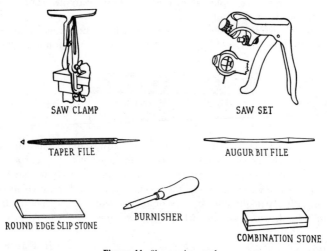

SAW CLAMP

SAW SET

TAPER FILE

AUGUR BIT FILE

ROUND EDGE SLIP STONE

BURNISHER

COMBINATION STONE

**Figure 11.** Sharpening tools.

SAW FILES. Saw files are procurable in sizes designated as extra-slim taper, slim taper, and regular taper. They are made in various lengths. A saw file must be purchased on the basis of the type of saw to be filed. There are so many variations that no description is possible here, but a good hardwareman will know what size to recommend for a specific type of saw.

BIT FILES. For augur bits a special file is obtainable with "safe edges" (no teeth) to permit filing the cutting edges of the bit without injuring the other metal parts.

CARBORUNDUM AND OILSTONES. A carborundum stone is fast-cutting, having fine, medium, or coarse grits. Some carborundum stones are made in combination: coarse on one side and fine on the other. Oilstones, which are made in different degrees of hardness, are required to put a keen edge on a chisel or plane bit. No tool kit is considered complete without at least one combination carborundum stone and one oilstone.

SLIP STONE. A slip stone is required for sharpening gouges and carving tools. Various shapes, sizes, and fineness of grit are available. This tool is not a necessity unless one has a set of carving tools or a gouge or two in his tool kit.

BURNISHER. A burnisher is a round-edged tool used for turning the edge of a scraper blade in order to make it cut. A very small turn will result in a very thin shaving and give a smooth finish to the hardest of the hardwoods or to any cross-grained piece of lumber. The end of the tool is pointed sufficiently to enable the user to run it in the groove (actually almost too small to be seen distinctly) and turn the edge back a little, the effect of which is actually to resharpen the scraper blade.

## POWER MACHINERY

There are many types of woodworking power machinery, any one of which would be adaptable to numerous kinds of jobs and operations required to do fine cabinetwork. Only the basic operations which can be performed on a few of these machines will be described below. Power machinery should be seen and demonstrated before it is purchased. With the purchase of a machine will come detailed instructions on how to set up and operate it.

The discussion which follows will be limited to the vertical drill, band saw, jointer, table saw, and electric grinder.

*Vertical Drill.* (See Figure 12.) The vertical drill is probably the simplest of the machines listed to set up and operate. A vertical-motion revolving spindle or shaft is made to receive a bit designed for wood or metal. If metal is to be bored, the machine is provided with an adjustable speed arrangement to permit changing the rpm (revolutions per minute) to conform to the material being bored; hardwoods are bored at a slower speed than are softwoods, and metals much slower than hardwoods. The tool stock in the vertical spindle is made to receive various sizes of drills or bits.

When a wood bit is used, the operator must be sure to place a piece of lumber on the metal table top to receive the bit spur after it has passed through the piece of wood being bored. Otherwise the wood bit will be ruined immediately upon contact with the metal.

Material to be bored should first be clamped to position to prevent it from whirling "free" as the bit enters the material; this could be the cause of a serious injury.

***Band Saw.*** (See Figure 13.) A band saw permits the operator to saw curved lines and corners, as he would not be able to do with a handsaw. The band-saw blades are made in various widths, from $\frac{1}{8}$ inch, for very fine scrollwork, up to 1 or $1\frac{1}{4}$ inches, for very heavy cutting. A small workshop band saw rarely has the power for a blade much wider than $\frac{1}{2}$ inch.

**Figure 12.** Vertical drill.          **Figure 13.**   Band saw.

Band sawing is a skill easily learned with a little practice. Forcing the blade around short curves or backing up the cut after the saw has entered the wood will result in either breaking the blade or pulling it off the wheels on which it travels.

Extreme care must be taken so that no part of the hand is ever allowed to come in line with the cutting edge of the saw when it is in motion.

Repetitive cutting on thin lumber is often accomplished by simply tacking several pieces together prior to band sawing. The only caution needed in this case is to be sure the nails are placed so that they will not be in the cutting line.

It is very important that the top and bottom saw guides be in the proper position to provide a smooth track or slot through which the saw blade can pass freely.

Tilting the top wheel by means of a tilting adjustment will cause the saw blade to move slightly across the edge of the rubber-tired wheel and thus make the wheel run properly in the saw guides.

*Jointer.* (See Figure 14.) The jointer, as its name implies, is used to joint the edge of boards to make them straight. The tilting fence permits a board edge to be beveled to any desired angle.

**Figure 14.** Jointer.

A skilled operator can also do tapered work on this machine. Tapering is the process of making a piece of lumber narrower at one end than at the other, as might be necessary in making a table leg, for example.

The jointer is recognized as one of the most hazardous of the wood-working machines. An expert operator will never let either hand pass directly over the revolving knives but will utilize such safety devices as pusher sticks or the metal guards which cover the part of the knife head not being used.

If a piece of lumber is passed too fast over the cutting head, the resulting cut will show many "wrinkles." Hence, to secure a smooth cut, the material should be moved slowly and carefully each time a cut is made.

*Table Saw.* (Figure 15.) A table saw for the average workshop need not be a large machine. The revolving spindle, on which the saw blade is placed, permits the use of either circular crosscut or circular ripsaw blades.

A saw blade should not project more than ¼ inch above the stock being cut.

Hardwoods must be fed slowly into the ripsaw; otherwise the blade will get very hot and burn, thus drawing the temper from the steel.

Pusher sticks are a "must" for all ripping operations, except when a board is of sufficient width to permit the right hand to be a safe distance from the revolving saw blade.

**Figure 15.** Table saw.

The saw fence, or guide, furnished with power table saws is adjustable to permit bevel cutting when ripping. Likewise, the crosscut fence can be adjusted to cut a variety of angles.

Saw blades are procurable in various sizes and shapes of teeth. Some teeth are made for ripping only, other teeth for crosscut work only; a third kind, known as a combination saw, may be used for either ripping or crosscutting.

The amount of work to be done on the table saw determines how many and what kind of saw blades to purchase. It is always an advantage to have additional sharp blades in reserve.

Groove or dado cutting is possible on most table saws. In a standard cabinet shop, where production time is an important factor, dado heads are purchased as part of the regular equipment of the shop. In a small shop, where time is not important, grooves and dadoes are often cut by making a series of saw cuts; the saw guide, or fence, is moved slightly each time a cut is made, and this process is continued until the desired groove width is reached.

*Electric Grinder.* (See Figure 16.) There are two kinds of electric grinder: the pedestal type, which is fastened to the floor, and the bench type, which is fastened to a suitable shelf or bench top. Bench grinders have a horizontal spindle with a thread and nut on each end to receive the grinding wheel. One wheel is purchased with a coarse grit for rough grinding, the other wheel with a very fine grit for fine tool grinding. The wheels also come in varying widths. The carpenter will normally use wheels at least 1 inch or $1\frac{1}{4}$ inches in width.

The bench grinder, as well as the pedestal type, is furnished with a tool rest and glass safety guard.

**Figure 16.** Bench grinder.

## CARE OF HAND TOOLS AND MACHINES

The life of a hand tool, a hand-powered tool, or a piece of powered machinery depends a lot on the care taken to keep the tool in good working order. Obviously, all metal tools should be stored or placed where they are not likely to be exposed to moisture. If a tool does become wet, or even damp, it should be wiped immediately and oiled to prevent rusting.

Since an orderly arrangement of a toolbox or a workshop means much in terms of efficient work and quality performance, too much time cannot be spent in carefully planning for the placing and storing of woodworking equipment.

A few simple rules will suffice to cover the care of equipment:

1. Provide a place for everything and keep everything in its place.

2. See that all tools are carefully protected when stored.

3. When using a plane, be sure to lay it down on its side when it is not in use; this will prevent the sharp plane bit from coming in contact with material which might ruin the sharp edge and thus will eliminate the necessity of a grinding and oilstone sharpening to put the plane back into cutting condition.

4. Avoid getting moisture on tools. On outside jobs this is sometimes impossible, and it is then essential to wipe and oil the tools carefully after using them. This precaution applies particularly to steel tapelines, which often come in contact with the ground or grass.

5. Avoid loaning tools. Loaned tools have a way of disappearing.

6. Put tools away in their proper place after using; they will then be ready the next time.

7. Keep tools locked up. This is good insurance against loss by theft.

8. Always buy good tools; they last longer and do better work. A good name-brand tool is always a safe buy.

## SAFETY RULES

There are a few fundamental safety rules that should be observed by any woodworker who uses hand tools and power machinery. These rules can be classified into four main divisions as follows:

*General:*

1. Provide a place for everything, and then keep everything in its place. This will eliminate some of the causes of tripping and falling.

2. When lifting a heavy object, be sure to lift with your legs, not your back, keeping your arms and body as nearly straight as possible.

3. Do not place objects on high, narrow shelves, or in window sills, or on top of stepladders, where they may fall and not only become damaged themselves but cause personal injuries as well.

4. Keep the floor free from oil and water.

5. Keep all working areas free from debris, lumber scraps, and tools.

6. Remove or bend down all protruding nails.

7. Keep all ladders in good repair. Inspect occasionally for weak rungs or steps. Never paint a ladder since you might thus conceal defects in the lumber.

8. Be sure, when using a ladder, that the bottom rests on a solid foot-

ing. This will prevent the ladder from slipping. Also, if the ladder is set too nearly in the vertical plane, it may fall out from the wall when the weight of the body is placed on it.

9. If any slivers of wood puncture the skin, remove them immediately to prevent infection.

### Clothing:

1. Loose clothing, ragged sleeves, and long neckties invite accidents. Sleeves should be rolled up neatly; this is particularly true when power machinery is operated.

2. Keep the soles of the shoes in good repair; thin soles are easily punctured with sharp objects, such as nail points.

### Hand-tool Usage:

1. When using sharp-edged tools, cut *away* from the body. A slip of the tool can cause a serious cut.

2. Keep all tools clean and sharp. A dull tool will slip much quicker than a sharp one, because of the extra pressure that may need to be applied.

3. Install a rack over the workbench to hold sharp-edged tools. This eliminates the hazard caused by leaving such tools lying on the bench.

4. Never use a file unless it is placed in a handle, which need be only a 1- by 1- by 6-inch piece of wood. If the sharp end of a file is left unprotected, it can cause serious injuries to the hand. Also, a handle simplifies the use of the file.

5. Keep all heads of metal tools tightly fastened to their wood handles.

6. The end of a metal tool—of a cold chisel, for example—will mushroom from constant hammering. The rough metal edge, which will slowly "grow" from continuous use, should be ground off before it becomes too large; otherwise a bit of this "extra" metal may fly off and strike someone, possibly in the face or eye.

7. Striking metal tools or any other hardened metal with a hard-faced hammer will cause chips of metal to break loose and fly; hence this practice should be avoided.

### Power-machine Usage:

1. Never oil or clean a machine while it is in motion.

2. Do not talk to the person operating a machine. Distraction from the operation being performed may result in serious injury.

3. Remove all tools and other articles from the top of a machine before turning on the power.

4. Check the adjustable parts of a machine before starting the motor; they may have become loose and need tightening.

5. Safety guards are made for the protection of the machine operator. Use them.

6. Before a machine is repaired, it is good safety practice to remove the fuses of the power switch; this will avoid accidents caused by another person's accidentally turning on the switch while repairs are under way.

7. When using a power saw, keep the hands away from the direction of travel of the saw. Also, never attempt to cut more than one piece of lumber at one time. Duplicate cutting can be done safely only if several pieces are tacked together so that they may be handled as one solid piece; the nails must be carefully placed to avoid cutting them.

8. Never lift a piece of stock over a machine that is running.

9. If it is necessary to clean off the top of a power saw while the machine is running, do so by means of a piece of stock; do *not* use your hands.

10. Use a pusher stick for small and narrow pieces of stock when operating the power ripsaw or the power jointer.

11. When operating a band saw, be sure that the height of the saw guide is adjusted before turning on the power. The machine should be stopped if the upper guide wheel needs to be raised or lowered to adjust the tension of the saw blade.

12. Cutting cylindrical stock on a band saw is not a safe operation; hence it should be done only when absolutely necessary and only by an experienced operator.

13. Glass guards, which are usually provided on a power grinder, and a pair of safety goggles should be used for all grinding operations.

14. The tool rest on the grinder should be properly secured and set close to the wheel. The tool to be ground should be held firmly on the tool rest.

15. Never grind small pieces on a grinder unless they are held in a proper holder.

Note: Figures 12 to 16 are supplied through the courtesy of Delta Power Tool Division, Rockwell Manufacturing Co., Pittsburgh, Pa.

# CHAPTER FOUR

# Joints

When a project is to be constructed from lumber, there are many different methods that can be used to join the several parts. This chapter outlines how to determine what type of joint to use and describes how each joint is made, as well as its particular application. The type of joint (or joints) selected is dependent on a number of factors:

1. The purpose of the job. For example, to serve only as ornament, such as a set of light hanging shelves; or to carry a load, such as a header over a wide window opening in a framed wall; or to join two pieces of milled lumber at right angles, in casing a doorjamb.

2. The material to be used—its working qualities and strength. Softwoods may require a different type of joint than do hardwoods.

3. How the job is to be used. Is the job part of a whole that will become a fixed part, such as a framed opening in a wall, or will it be moved occasionally, such as a piece of furniture?

4. Method of fastening the joint, which may require nails, screws, glue, dowels, or metal fasteners—or possibly a combination of several of these.

5. The angle of the grain of one piece of lumber in relation to the grain of the piece to be joined to it

**Note:** The time involved in making the joint is *not* a factor that needs consideration in this discussion. In a production cabinet shop, joints are selected on the basis of ease of manufacture and durability. It is assumed that time is not an important element; hence the best methods of construction will be used.

## TYPES OF JOINTS

The types of joints illustrated in Figures 1 to 10 will each be discussed in terms of its purpose, where it is best used, how it is made, and how it is fastened. These joints include butt, scabbed, cleated, half-lapped, miter, dowel, dado and rabbet, tongue-and-groove, coped, and scribed.

*Butt Joint.* (See Figure 1.) The butt joint is the simplest of all; in house framing it is a good "load-carrying" joint. It is very easily made: the end of a board or framing member is cut off perfectly square and then "butted" against a piece of lumber having similar dimensions.

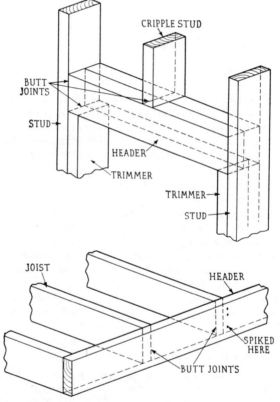

**Figure 1.** Butt joint.

Nails are driven through the face of one piece of stock into the cut end of the second piece. The diameter and length of the nail are determined by the size of the lumber being used. In framing floor joists, for example, a 16d box, 16d common, or 20d common nail is often used.

The butt joint is an excellent joint for rough-carpentry jobs, where strength is most important and appearance is not a prime requisite.

*Scabbed Joint.* (See Figure 2.) Sometimes it is necessary to join two short pieces of lumber to make a longer piece; the scab "ties" the two pieces together. Definitely a rough-carpentry joint, it is used only where appearance does not count or where the joint may eventually be completely hidden. The piece of lumber, or "scab," is usually face-nailed to the lumber to be joined; common or box nails of suitable length are used. The diameter or gauge of the nail must be carefully selected; a nail with too large a diameter will split the scab, thus causing the joint to lose its effectiveness and strength.

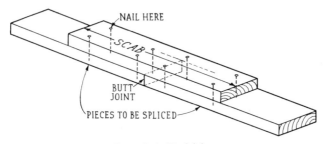

**Figure 2.** Scabbed joint.

*Cleated Joint.* (See Figure 3.) The cleated joint has some of the characteristics of the scabbed joint; it is, however, a carpentry semifinish method, used to fasten two (or more) boards together to make a wider board. It is often used in making rough doors for farm buildings.

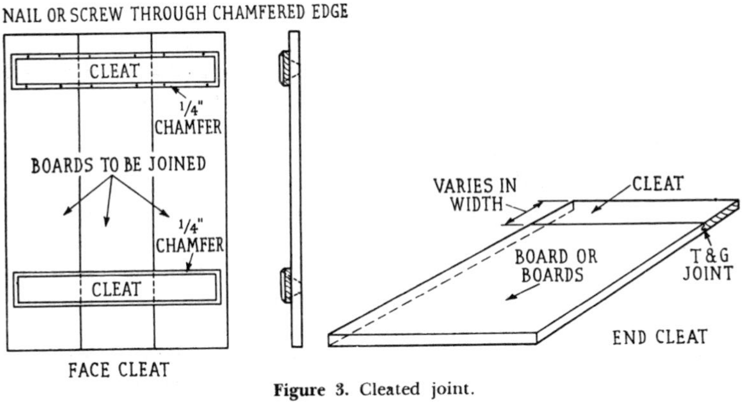

NAIL OR SCREW THROUGH CHAMFERED EDGE

CLEAT

1/4" CHAMFER

BOARDS TO BE JOINED

1/4" CHAMFER

CLEAT

FACE CLEAT

VARIES IN WIDTH

CLEAT

BOARD OR BOARDS

T & G JOINT

END CLEAT

**Figure 3.** Cleated joint.

*Face* cleats are fastened on with screws or nails. The edges of each cleat are chamfered on all four edges; screws or nails are then driven at right angles to the face of the chamfer. To make a neat-looking job, carefully space the nails or screws. The size of nail or screw used is determined by the thickness of the cleat and the material which is to be cleated. Obviously, no fastener should be used that will project through the opposite face of the boards.

Cleats should be placed down from the top and up from the bottom edges of a door to give a pleasing appearance. The distance from the top edge of the door to the top cleat is usually less than the corresponding distance at the bottom end of the door.

An end cleat is used when making a pastry board or drawing board. The tongue-and-groove joint (see Figure 8) is most often used. A board that has been end-cleated will not split.

***Half-lap Joint.*** (See Figure 4.) A half-lap joint is useful in constructing a workbench, a gate, or a rough door when exposure to the weather must be considered. It is used primarily to provide rigidity and strength

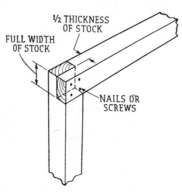

when two pieces are to be joined, usually at a 90-degree angle. One-half of the width and thickness of the stock is removed on the end of each piece of stock. This is done by sawing, preferably, or chiseling, or both. Care must be used to be sure that the stock removed is from the *same* side of each piece.

An accurate job will result if the joint is laid out carefully by marking *exactly* in the center of the stock and cutting down from each end an amount equal to the exact width of the stock. The flat bottom

**Figure 4.** Half-lap joint.

of each cut must be exactly parallel with the face of the stock; otherwise the final job will be twisted.

The joint formed when the two half laps are fitted together is held by means of nails (in very rough carpentry) or, preferably, by countersunk wood screws, at least two to each corner joint. To increase the strength of the joint, use glue and C clamps, and no screws; or, to make the job ready for immediate use, use glue with screws.

***Miter Joint*** (See Figure 5.) This is a finish joint and is not used where strength is the main factor. It is most often used to make the 90-degree angle formed by the top and side casings of a window or doorframe and to construct picture frames.

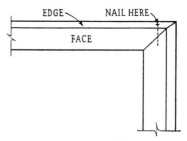

Figure 5. Miter joint.

If the pieces to be mitered are molded, the 45-degree angle of the cut must be very carefully laid out; otherwise the arrises (corners) and curves will not exactly coincide.

For repetitive cuts a wood or metal miter box should be made. A backsaw (see Figure 4, Chapter 3) is best for cutting, and a sharp block plane will give the cut a very smooth finish, thus assuring a perfect joint. This joint permits nailing in two directions to prevent the joint from "opening up."

*Dowel Joint.* (See Figure 6.) Doweling stock is made from birch or maple in several different diameters, such as $\frac{1}{4}$-inch, $\frac{3}{8}$-inch, $\frac{1}{2}$-inch, $\frac{3}{4}$-inch, and 1-inch. The standard length of each piece is 36 inches, but doweling stock can be purchased in short lengths, ready for use.

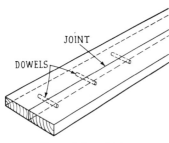

Figure 6. Dowel joint.

The dowel joint is used for fine cabinetwork, such as a cabinet door, to ensure strength and holding power where the rails and stiles are joined; or for joining stock to make table tops or wide shelves.

The length of each dowel used in door construction depends on the width of the door stile. For gluing shelving together, the dowels are usually 3 to 4 inches in length.

Preparing a dowel joint is a good test of craftsmanship. The holes made to receive the dowels must be carefully bored at *right angles* to the joint edge and *parallel* to the surface of the board.

The augur bit must be exactly the correct size; otherwise the dowel will either not fit the hole or be too loose and hence of no value whatever.

Dowel holes should be carefully reamed with a countersink. Beveling the edge of the dowel hole provides space to receive an extra amount of glue. The ends of each dowel should also be beveled slightly, using a dowel pointer. This operation makes it easier to put the dowel in the hole. Dowel holes should be bored slightly deeper than one-half the length of the dowel. An open joint will result if the dowel pins are too long.

In joining two boards, the end dowels are kept at least 2 inches from each end of the board. Intermediate dowels are spaced evenly. There is no set rule about the number of dowels to use; probably one dowel should be placed for each 10 inches of length.

When gluing the joint, it is necessary to put glue in *each* dowel hole as well as on the dowels. On a tight-fitting dowel all the glue will be rubbed off as it enters the dowel hole, and the joint will lose its effectiveness and strength.

Hot glue is preferable on the dowel joint, provided that one has a sufficient number of glue clamps, which should have been previously adjusted to the width of the stock to be glued. Also, there should be no draft while the job is being done; cold air will make hot glue set very quickly. Cold glue, however, is very satisfactory in regard to strength; it is also easier to work with, since setting time is not an important factor.

Excess glue should be removed with a damp rag after a job is in the glue clamps. This makes the cleaning and sanding job easier.

***Dado and Rabbet Joints.*** (See Figure 7.) A dado joint and a rabbeted joint are very similar in shape and purpose and are usually made with the same tools. The rabbet joint is used at the corners of a window or doorframe or in making a set of shelves. The dado joint, which is a groove made at right angles to the grain of a board, is used to hold a shelf in exact position to the vertical supporting member (called a jamb or cabinet end).

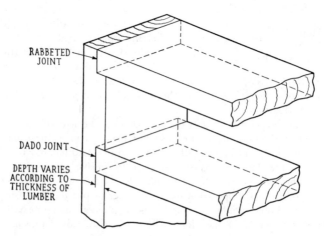

**Figure 7.** Dado and rabbet joints.

Rabbets and dadoes are made in varying widths to suit the stock which is to be fitted into them. A ¾-inch shelf would require a ¾-inch dado; if the stock is $1\frac{3}{16}$ inches thick, then the dado will need to be exactly $1\frac{3}{16}$ inches in width. Likewise, the rabbet at the corner of a window or doorframe is made to fit the material being used.

The rabbet joint permits nailing in two directions to increase the holding power of the joint. In the dado joint, nails, if used, are needed only through the end of the shelf; the groove itself provides the necessary strength to hold the piece of material to its exact position and obviates the need for a shelf cleat, often used in very rough carpentry jobs in place of a dado.

Dado and rabbet joints can be made by hand or on a table saw. The directions for making a dado on a table machine are simple: put on the dado head, which has been "built up" to the required thickness; set the fence to the proper dimension; be sure the dado head projects above the table exactly the desired depth (make a trial cut on a scrap piece of lumber); then cut each dado at a predetermined location. The layout work for each dado should be done with a rule and pocketknife to assure exact location marks.

To cut a dado by hand requires a backsaw, chisel, and router plane, in addition to the usual layout tools such as a rule and knife.

After two marks have been made, locating each dado, clamp the board to the bench top; a C clamp or hand screw is very useful for this purpose. Select a straight piece of thin finish stock of suitable length and width (¼ inch by 1½ inches by length of dado cut) and tack this piece onto the stock to coincide exactly with one of the dado lines. With this guide strip in place, use a backsaw and carefully make a saw cut to the required depth. Repeat this process for the second line of the dado cut.

Next, roughly chisel out the wood between the two saw cuts, taking care not to go deeper than the depth of the finished dado cut. Complete the cut by using the router plane, which has been adjusted to cut the exact depth of the dado. This tool will give a very smooth appearance to the bottom of the dado and, of equal importance, give assurance that the dado is exactly the correct depth in its entire length.

A rabbet joint is made exactly as described above for the dado, except that the cut is made at the end of the stock. If the board to be rabbeted is true and straight, a rabbet can be made by using a saw across the grain and then making a series of short strokes with a wide chisel and removing the wood. On the power saw a rabbet can be cut in both directions: first, by passing the board across the saw and, second, by holding the board vertically and passing the end of the board across the saw.

Dado and rabbet joints are commonly used on many cabinet and other finish-carpentry jobs. Their ultimate strength lies in the accuracy of the layout skill and cutting performance. Loose-fitting dadoes and rabbets will give neither a strong joint nor a finished appearance. A craftsman who is working on a job that is new to him will therefore find it worthwhile to make a few "practice" dadoes and rabbets to acquire the skill necessary to execute a finished job.

*Tongue-and-groove Joint.* (See Figure 8.) A tongue-and-grove joint (T-and-G for short) is usually a machine-made joint and is used primarily in flooring, wall sheathing, finishing the face of a wall with knotty-pine boards, and doing any similar work in which a number of boards must be joined to make a smooth, flat surface. A groove is made in the edge of one board to receive a tongue made in the edge of another board. Each piece of T-and-G stock will therefore have one edge grooved and the other edge tongued.

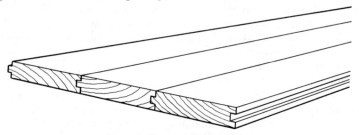

**Figure 8.** Tongue-and-groove joint.

No instructions will be given on how to make a T-and-G joint, for it is primarily a machine-made joint, and T-and-G stock can be purchased at any lumberyard ready for immediate use.

Extreme care must be used when nailing T-and-G material to its place. Casing nails are used for flooring and are driven into the tongue edge at an angle. Two difficulties usually arise: (1) the tongue gets battered, so that it is quite difficult to get the groove of the next board to fit tightly over the tongue unless all "overwood" is removed; (2) the nailer bruises and mars the top edge of the flooring; when the next piece is fitted tightly to it, the marred edges plainly show. The second difficulty is overcome only by careful nailing: the hammer should be held *almost vertically for the last blow,* so that the hammer head hits the lower edge of the nail, thus driving it farther in without bruising the edge of the material.

*Coped Joint.* (See Figure 9.) A coped joint is used when fitting one piece of molding at right angles to a second piece. This joint is most

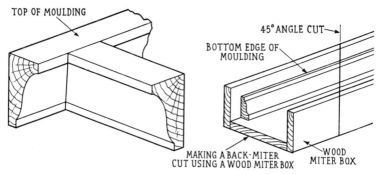

Figure 9. Coped joint.

often used when fitting molding on the walls of a room or on the surface of a piece of paneling.

A coped joint is not difficult to make, provided that the cabinetmaker knows how to make a "back miter." The molding is placed in an upside-down position in the miter box, and a miter cut is made as though the joint were going to be an inside miter cut. A coping saw is used to cut along the profile of the molding formed by the miter cut. This makes a cut on the end of the molding exactly the same shape or profile as the molding face to which it is to be fitted if the coping-saw cutting is done exactly on the profile line and at right angles to the face of the molding.

*Scribed Joint.* (See Figure 10.) A scribed joint is required when a piece of finish stock is to be fitted at right angles against a slightly uneven surface, such as a plastered wall. For example, if a hall is not wide enough to have full-width door casings on the doors placed at the ends of the hall, the door casings must be scribed to fit the wall against which they are fitted. The scribers (see Figure 2, Chapter 3) simplify the process of making the scribing line which indicates where the casing must be cut.

It is important to note that a piece of lumber, when being prepared or placed in posi-

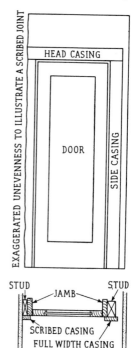

Figure 10. Scribed joint.

tion for scribing, must be located so that the finished edge (the edge that is not cut) is exactly parallel to the edge to which it is to be nailed. For example, in scribing a piece of baseboard to fit a floor surface, the top edge of the baseboard must be exactly parallel to the floor surface.

A scribed joint is always "undercut" to allow for a corner which may not be exactly a right angle. This situation can easily be detected by checking a plastered inside angle with a try square. The "offsquareness" may be only a trifle; undercutting the scribed joint will assure a finish joint.

The pencil point in the scriber must be very sharp and the tool kept exactly level (or vertical) while the scribe line is made. Twisting the scriber will not give a true scribe line; hence the piece of stock will not fit the uneven surface.

# CHAPTER FIVE

# The Foundation

In house construction a most important job is building the foundation. Any discrepancy in measurements, any "out-of-square" or "out-of-level" measurements, will show up in every succeeding job that is constructed on top of the foundation. Hence, too much caution cannot be taken when constructing a house foundation. It is also true that the various jobs required to construct a foundation are complicated by the fact that there are no guide points from which to take starting measurements, other than property stakes designating the lot corners. The job is particularly difficult when the ground is not level, that is, when the house is to be built on a hillside.

The following pages will cover, in detail, the various jobs done in house-foundation work including batter-board layout, trench excavating, concrete-form construction, and backfills. No description will be given of brickwork, flat concrete work, or concrete steps since these are not carpentry jobs.

## HOUSE LAYOUT

*Building-layout lines.* Building-layout lines are usually placed to represent the outside face of a building, although the foundation footing extends beyond the face of the foundation wall, as it is several inches wider than the wall thickness. To illustrate, a foundation wall 6 inches thick will require a footing 12 inches wide so that, in this example, the outside of the foundation footing would be 3 inches beyond the building line. For an 8-inch wall the footing would be 4 inches beyond the face of the wall. (See Figure 10.)

It is a simple matter to locate the edge of the footing by measuring 3 inches (or whatever the amount may be) outside the building line and then to construct the footing form at that place.

*Transit.* Modern contractors use the transit level for their layout work, by means of which building lines can be established which are level and also form perfect right angles. A man who desires to do his own layout work can do an accurate job without a transit, but obviously the transit is the more efficient method when time is an important factor.

Carpentry layout tools consist of a good 24- or 30-inch level, a 50- or 75-foot tape, a steel square, a hand ax, an 8-point crosscut saw, a claw hammer, a plumb bob, and a 10- or 12-foot straightedge. It is often possible to get accurate results by using a straight piece of stock as a straightedge. This straightedge should not be confused with the one described in Chapter 3, page 25, which is used primarily for finish work. The description below assumes that the carpenter will have available a pile of lumber from which a satisfactory piece can be selected.

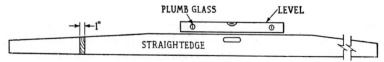

**Figure 1.** A straightedge.

*Straightedge.* A straightedge for building layout (see Figure 1) is made as follows:

1. Carefully select a 12-foot piece of 1 by 6 stock with good grain that has one straight edge, which is ascertained by "sighting" along the edge.

2. Locate the center of the board and then plane a 4-foot section on the opposite edge of the board, making it exactly parallel to the straight edge.

3. To lighten the tool, taper the remaining top edge of the board so that the ends are roughly 3 inches in width.

4. Bore several 1-inch holes near the top edge; chisel and rasp out to make a smooth handhole, by which the tool may be carried.

The "straightedge" is now ready for use along with the carpenter's level.

*Locating House Corners.* (See Figure 2.) Before constructing the batter boards that hold the lines which show the exact shape of the building (see below), single stakes are driven giving the *approximate* location of each of the four corners of the building. (The batter boards are then constructed around each stake.) These stakes are located by the use of a tapeline. Measurements are usually made from the front-sidewalk line, assuming that the house is to be constructed parallel to the sidewalk. City ordinances or specifications must be read to find out the minimum allowable setback distance, often 25 or 30 feet.

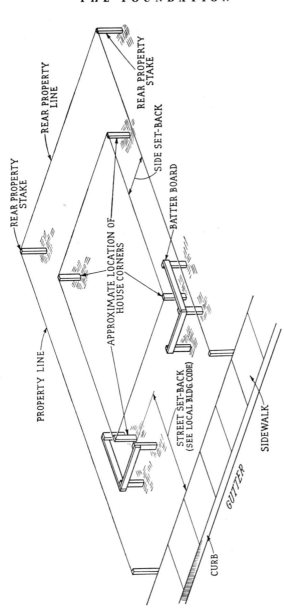

**Figure 2.** Locating house corners.

The distance the house is to be constructed from the side line of the lot is shown on the blueprint or otherwise determined by the owner. Building ordinances, which in many cities are quite specific about the minimum distances a house must be set in from the side line, must be checked.

The property stakes are important at this point; they must first be located according to a city map of the lot, or if no property stakes can be found, the services of a surveyor may be required to determine the *exact* location of the property lines. Ordinarily there should be one property stake for each lot corner.

The steps to follow when locating the *approximate* house corners are:

1. Measure from the front-sidewalk line the setback distance and drive a stake near each side line of the property.

2. Stretch a strong piece of building twine from stake to stake. This establishes the approximate location of the front line of the building.

3. Stretch a line from a property stake in the front of the lot to a corresponding property stake at the rear end of the lot. This represents the side limits of the lot.

4. Measure "in" from this line the building setback, drive two stakes, and stretch a line from one to the other. This represents one of the side building lines.

5. Measure from the front line a distance representing the dimensions of the house from front to rear; drive two stakes and stretch a line to establish the rear line.

6. Repeat this step for the remaining side line, using the side line first established (step 4 above).

7. Where the lines cross at each corner, drive a stake. This gives the *approximate* location of the four corners of the house. A plumb bob

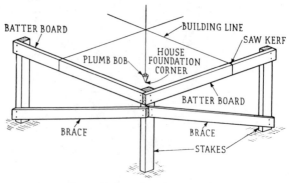

**Figure 3.** Batter-board construction.

dropped at the line intersections will give an accurate location for the stake (see Figure 3).

8. Remove all building lines. You are now ready to erect the batter boards around each corner stake. On these batter boards the *exact* corner locations will be marked.

**Note:** Architects may design a house with the first floor only a few inches above the ground which surrounds it, requiring that excavation work be done to provide for girder and post construction and for crawl space under the joists. Building codes are quite specific as to the minimum amount of space—18 inches is standard—which must be left below the joists.

It may be necessary to excavate a considerable amount of dirt. In addition to the crawl space there may be a basement. The excavation required under the joists is done after the house is "roughly" laid out and prior to doing any exact batter-board construction. Basement walls, however, must be located and laid out accurately. This is done after the batter boards are erected.

*Constructing Batter Boards.* Batter boards consist of three stakes and two horizontal members placed at least 3 feet outside the building line (see Figure 3). The stakes need to be carefully pointed, since a stake with an uneven point will invariably twist (see Figure 4), and driven in sufficiently to make them rigid. Horizontal members must be very carefully placed in position, using the carpenter's level to make them level.

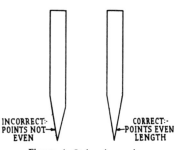

INCORRECT:- POINTS NOT EVEN

CORRECT:- POINTS EVEN LENGTH

**Figure 4.** Stake sharpening.

The exact height of a batter board is not too important, provided that it is made a little higher than the top of the finished foundation. It is very easy, once the building lines are pulled taut, to measure *down* any desired distance. If the lines are too low, they will be in the way when constructing the concrete forms.

**Note:** Some carpenters prefer to construct batter boards the exact height of the top of the finished foundation wall to simplify the work of constructing forms at the correct height. If this procedure is followed, it is necessary, before any batter board is constructed, to determine what the foundation height above grade line is to be. Such problems as the slope of the front yard toward the street, the direction of the slope of the street, the number of steps desired from the grade line to the finished floor of the house, and the levelness of the lot itself must all be considered prior to erecting the batter boards.

If a blueprint of the house is available, the problem is quite simple, for in this case the drainage and slope problems have been studied and the solution determined by the architect; hence grade and foundation heights can be found on the blueprints.

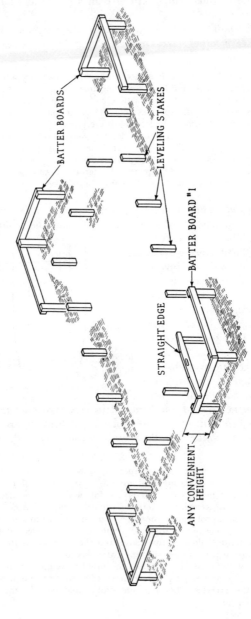

**Figure 5.** Using straightedge to get foundation heights level.

Each batter board should be level with the other three; this is accomplished by the use of the straightedge and level (if a transit is not available).

Obviously a 12-foot straightedge would not be long enough to reach from one corner of a building to the next. Hence, after the first batter board is erected, a series of stakes are driven at 12-foot intervals around the building (see Figure 5). One end of the straightedge is placed on a stake, the level is checked for exact levelness, and a second stake is driven. Each time a new stake is driven *the straightedge and the level must be reversed end for end.* This simple reversing procedure will ensure close accuracy in establishing the height of each corner batter board. Any slight variation in the accuracy of the level or straightedge is automatically adjusted by the reversal of the straightedge and level. The final test of the job is to see whether the straightedge is level when it is rested on the first batter board from which the layout work was started.

If the work has proceeded as described, there should be four batter boards erected in readiness to receive the building twine. Inside angles or other breaks in the building line will require similar batter boards. On some buildings which may have many angles, it is good practice to erect what amounts to a continuous batter board all around the building. The various corners and angles can then be easily located at any required point on the top edge of the continuous batter board.

*Locating Exact Building Dimensions.* Saw kerfs are made at an exact location in the top edge of the horizontal members of the batter boards to hold the building twine. The procedure is outlined below:

1. Accurately measure the distance from the front property line to the side piece of each front batter board; mark for and make a saw cut.

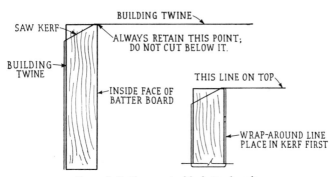

**Figure 6.** Cutting saw kerf in batter boards.

This cut is made at an angle so that the twine will always be even with
the top edge of the board (see Figure 6).

2. Locate one of the side lines in a similar manner, measuring from
the side property line.

3. Check these two lines for right-angle accuracy by using the 6-8-10
method or a multiple thereof. This is done by measuring, say, 6 feet (or
12 feet) on the front line and placing a common pin in the line at this
point; do the same for the side line, only use 8 feet (or 16 feet). Be sure
that lines are taut before placing pins. Then measure carefully from
pin to pin. If the distance is 10 feet (or 20 feet), the angle is perfectly
square (see Figure 7). If there is any variation, the rear end of the side
line must be adjusted accordingly, the line restretched, the pin reset,
and the diagonal measurement taken again. This process is continued
until the 10-foot (or 20-foot) measurement from pin to pin is registered
on the tapeline.

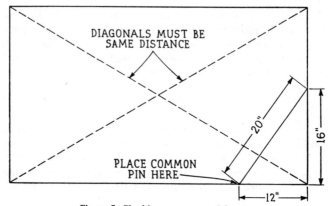

DIAGONALS MUST BE
SAME DISTANCE

20"

16"

PLACE COMMON
PIN HERE

12"

**Figure 7.** Checking squareness of layout.

**Note:** The geometrical principle involved in this method is simple: in any
right-angled triangle the square of the hypotenuse equals the sum of the
squares of the other two sides ($a^2 + b^2 = c^2$). Any right triangle that measures
3 feet on one leg and 4 feet on the second leg, therefore, will have a 5-foot
hypotenuse (see Figure 8). Any multiple of these numbers will also do; instead of
3-4-5, use 6-8-10 or 12-16-20, etc.

4. Measure from the front line and locate the saw cuts for the rear
line of the building.

5. Measure from the side line and locate the other side-line cuts.

6. Place the building twine in position.

**Note:** To hold the line taut in the saw kerf: (1) wrap the end of the twine
around the batter board and place in the saw kerf; (2) let the horizontal line
rest *on top* of the "wrap-around line." The harder you pull, the tighter becomes
the holding power of the line in the saw kerf (see Figure 6).

7. As a final test for squareness, check the diagonals of the rectangle or square formed by the building twine. If the dimensions are the same, the building layout has been accurately done. If the measurements are different, then adjustments must be made until the diagonal measurements are the same (see Figure 7).

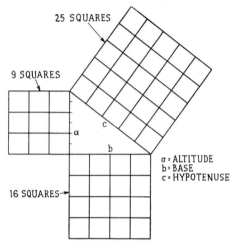

**Figure 8.** The geometrical principle of $a^2 + b^2 = c^2$ demonstrated.

8. Now "sight" the building lines for levelness; all lines should be in perfect alignment if the straightedge and level work have been accurately done.

9. Any inside angle or wall offset is laid out in the same manner, using the basic outside building lines as points of measurement.

## EXCAVATING FOOTING TRENCHES

Footing trenches are made to conform to the width of the footing (see Figure 9). The depth measurement will vary; it should be at least 6 inches below the established grade line. In every case it is essential to extend the foundation below possible frost penetration, even though firm bearing soil is found at a shallower depth. The foundation will then not be upheaved by freezing. In some districts frost penetrates as much as 6 feet. The local building code or the common practice in the community is the best guide to the depth of a footing below the grade line. The blueprints will indicate exact depth measurements for basement walls.

On a lot that slopes, the trenches are often stepped down at stated intervals; the wall on top of the footing is then increased in height so that a level line is maintained at the top.

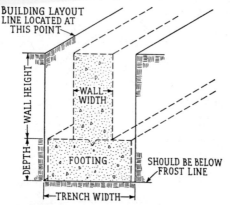

**Figure 9.** Excavating footing trenches.

Trenches should be carefully dug. When soil conditions permit, it is sometimes possible to make the walls of the trenches serve as a form for the concrete. A very loose soil, however, will obviously require concrete forms for the footings.

If filled earth is encountered, the trench must be dug until a solid soil is found. This is a building-ordinance requirement which must be followed, for the carrying load of a foundation is dependent on the condition of the soil on which it is built.

A square-pointed shovel is best for trench digging as it enables the user to make vertical and straight earthwalls and a level floor in the trench.

Care should be taken to place the excess earth far enough away from the building so as not to interfere with the work of form construction. It is always best to place the earth outside the foundation rather than inside, where it would cause an uneven working surface for the carpenter as he works on the floor unit. Also, there is a code requirement indicating the amount of clearance below the floor joists. Piles of earth inside the foundation would therefore have to be moved.

It is good practice to cut a short stick the required width of the trench; this provides a simple measuring device to keep the trenches the correct width as the digging proceeds.

Building lines, if kept taut, can be used as measuring points to establish the trench-wall location and to keep the floor of the trench parallel with the top of the foundation wall. Slight variations in this vertical distance are not important as long as the minimum thickness of the concrete footing is obtainable.

## GENERAL INFORMATION ON FORM CONSTRUCTION

The topic of form construction is very comprehensive. The discussion below, however, will be limited to corner construction, straight-wall construction, inside angles, whalers and tie wires, bracing, reinforcing steel, and sill bolts. Brief information will be given on the need for oiling a form and the method of stripping a form.

*Lumber Sizes.* Form construction for a house foundation is not difficult. Foundation walls are rarely high, except in certain cases involving side-hill construction or high basement walls. It is customary to use 1 by 6 boards, S1S1E (surfaced one side and one edge) and 2 by 4 studs and whalers. Braces for low walls are often 1 by 4, but the carpenter will sometimes use short ends of 1 by 6 stock and split them to 1 by 3. Two form walls are required to make a concrete wall; the outside one is made first and then the inside one constructed and tied to the outside one, which has been rigidly braced.

Ordinarily concrete foundation walls are not plastered except on a building having a plastered exterior. If the face of the wall is to be plastered, the rough sides of the form boards are placed "in" to cause a rough surface on which the plaster will more readily adhere. For a finished concrete wall face, the material is placed surfaced side in. The cracks between the boards are made as tight as possible; hence the stock is ordered S1E to make it all the same width. On most construction jobs subfloor stock is used for form building; much of it can be recovered for subfloor material.

Note: In Class A construction, forms are often lined with 3-ply paneling.

Oiling the inside face of the forms is an excellent method to prevent the form boards from sticking to the concrete. Crude oil is excellent for this job.

*Reinforcing Steel.* (See Figure 10.) Reinforcing steel in the concrete footing is of considerable value in a house foundation, for it provides a continuous horizontal bond for all corners and angles. The use of steel will also prevent foundation settlement if an occasional short soft spot is encountered when digging the trenches. The size (diameter), number of bars, and the location of each bar will be shown in the blueprints. Building ordinances must also be checked, as the foundation requirements for steel given in a building code have been carefully engineered to give maximum results in terms of load-carrying capacity.

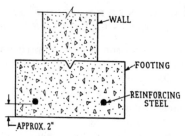

Figure 10. Reinforcing steel.

*Nail Sizes.* Nails for form-construction boards should not be larger than a 6d box, to facilitate the stripping of the forms after the concrete is poured. Whalers will require an 8d box, which is toenailed into each stud. Sometimes double-headed nails are used on form braces to permit easy removal.

*Studs.* Studs are normally placed 2′0″ on center (o.c.) as this measurement conforms to the even-foot length of lumber and avoids waste of lumber (see Figure 11).

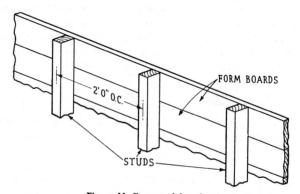

Figure 11. Form-stud location.

*Whalers.* (See Figure 12.) Whalers are usually 2 by 4 stock, placed horizontally against the top edges of the outside wall studs to hold the wall to a true straight surface and prevent the form from moving out of shape when the concrete is poured. A whaler must be carefully selected for straightness and strength; hence cross-grained or knotty lumber should not be used.

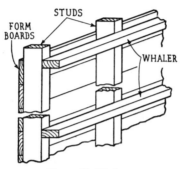

Figure 12. Whalers.

**Form Bracing.** (See Figure 13.) Form bracing is accomplished by driving a series of stakes a suitable distance outside the form to which 1 by 3 or 1 by 4 braces are nailed; the top end of the brace is nailed to the top of the form; the lower end is nailed at the ground line of the stakes. The number of braces depends on the height of the wall. The height of the wall determines the size of the brace stock; that is, a high wall would require 2 by 3 or 2 by 4 braces. It must be emphasized that it is much easier to brace a form properly, putting in more rather than fewer

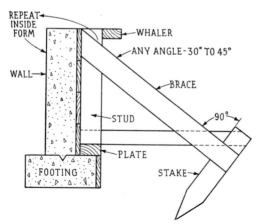

Figure 13. Form bracing (inside form braced similarly).

braces, than to build on a crooked concrete wall that cannot be straightened.

A skilled concrete-form man will obtain maximum strength and rigidity in his forms with a minimum of bracing; an inexperienced man will do well to be doubly sure his forms cannot get out of line while the concrete is being poured, for this operation puts a tremendous strain on a form.

*Straightening a Form.* After the outside form is completed and the corners and angles located and braced to conform exactly to the building-layout lines, the form is straightened by sighting along the top of the outside-form wall, moving the form until it is perfectly straight, and then holding it to place by means of angle braces. An even more accurate job is secured by establishing various intermediate points along a form wall to conform exactly with the building line, rather than depending on sight. The form is straightened before the whalers are placed in position. The whalers are added mainly to supply rigidity to the form, although they will, if selected for straightness, simplify the straightening process.

*Tie Wires and Spreaders.* (See Figure 14A.) Tie wires and spreaders are placed at suitable horizontal and vertical intervals (depending on

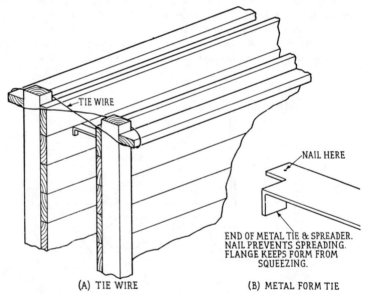

(A) TIE WIRE          (B) METAL FORM TIE

**Figure 14.** Form ties.

the height of the wall) to hold two wall forms exactly parallel. The wires are wrapped around opposing studs (including the whalers whenever possible). These tie wires are very important as they give rigidity to the forms. The wires become embedded in the concrete and, when the forms are removed, must be cut off with a suitable wire nipper. The size of the wire should be No. 12, 13, or 14. The larger the number, the larger the wire.

A piece of 1 by 1 scrap lumber (called a *spreader*), the length of which is the exact inside-dimension width of the form, is used along with each tie wire to keep the form from being pulled together when the wire is tightened around the whalers. The spreaders *must* be removed as the concrete is poured. If they are left embedded in the wall, the swelling of the wood will cause cracks in the concrete and produce an unsightly appearance.

Metal form ties are also available which act both as ties and as spreaders (see Figure 14*B*). While a form must be kept from spreading apart when the concrete is poured, it is also important that the correct dimension between the forms be maintained before the concrete is poured. A metal tie-spreader combination can be purchased for this purpose.

***Sill Bolts.*** (See Figure 15.) Sill bolts to receive the sill are placed in the top of the concrete and extend about 2½ inches above the concrete. The length of these bolts should be at least 8 inches; here again it is necessary to consult the building ordinance or the blueprints, which will usually show the required bolt size and the minimum center spacing.

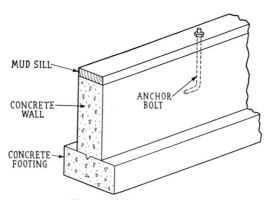

**Figure 15.** Sill bolt or anchor.

***Ventilation.*** Foundation areas under the first-floor joists must be properly ventilated to keep them from becoming musty. The number of ventilators required for a house foundation is specified in the building code, usually on the basis of "2 square feet of ventilator for every 25 linear feet of wall." No exact figure can be given as it may vary in different ordinances. The sizes of the vents are also stated in the code.

Vents must be placed high enough off the ground to prevent water seepage through them. A common practice is to secure ventilation by cutting holes in the floor joists, sometimes called rim joists, that are placed around the edge of the foundation. For a building using 2 by 6 floor joists a net space of $5\frac{1}{2}$ by 14 inches can be secured. One-quarter-inch mesh is nailed over this opening, or if a metal vent is used, the wire is already fastened to the metal edge on all sides of the vent. See Page 103.

Sometimes hollow tile is used for vents, in which case it is placed *inside* the form at the required height prior to pouring concrete. If the tile is not large enough to fill the space between the form boards (foundation-wall thickness), it is necessary to make a small wood frame of the same height and length as the tile to fill the remaining space. When this wood frame is removed after the forms are stripped, the result is a small reveal on the face of each ventilator (see Figure 16).

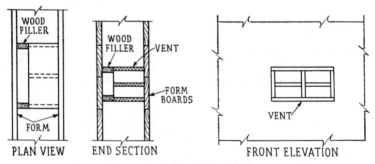

PLAN VIEW          END SECTION          FRONT ELEVATION

**Figure 16.** Setting hollow-tile vents.

*Form Stripping.* A suitable waiting period should occur before a concrete form is stripped. On large construction jobs building ordinances give the exact period for the concrete to set before the forms may be removed. For house construction no stated time is given; it should be obvious, however, that the longer the period of time between concrete pouring and form stripping, the harder the wall will be and the less likely to crack or to break along the edges. It is good practice, however, to remove the forms within 48 hours after pouring to permit the air to get to the wall faces.

*It is definitely bad practice* to pour a foundation one day and then strip it the next day and start construction, although this is done by commercial builders who are anxious to get the framing work started. Concrete takes time to set properly; in hot weather it should be sprayed to prevent it from drying too fast. It is recommended that at least four days should elapse before framing construction is started.

Form stripping is done, in a general sense, in exactly the reverse order of erection. Braces are first carefully removed. At no time should any prying be done on the wall itself. It should now be quite apparent that the reason for leaving nailheads protruding or for using double-headed nails is that this procedure permits nails to be withdrawn easily.

It is an excellent safety rule to provide suitable storage space for the forms as they are removed. The form stock should be piled in parallel fashion. Nail pulling is generally left until all forms are removed; this speeds up the operation of both stripping and cleaning. Short pieces of form lumber that have no further building use should be piled in a heap and burned to eliminate the hazard of protruding nails and sharp ends of split lumber.

***Summary of Form-construction Principles.*** The basic principles of form construction may be summarized as follows:

1. A form properly constructed will "stay put" during the pouring of the concrete. It is much easier to brace a form thoroughly than to wish you had after the concrete is poured.

2. The pouring of the concrete inside the forms actually tightens each board against the studding; hence a short nail is used, since it is needed only to hold the boards to their proper place and prevent warping.

3. All inside-angle studs must be planned for easy removal; nails are not driven in and, therefore, can be pulled out by means of a pinch bar.

4. Whaler stock must be selected carefully for straightness; a crooked foundation wall will cause considerable difficulty when the actual framing work begins.

5. All forms should be set exactly level; an "out-of-level" foundation causes endless difficulties which can never be completely corrected.

6. Once a form is finished it should be obvious that it should not be stepped on or otherwise abused.

7. In stripping a form, sufficient time should elapse to give the concrete time to become hard and set. No pressure should be placed on a "green" concrete wall; all form lumber should be piled neatly and all nails removed. Scrap lumber should be burned to eliminate the protruding-nail hazard.

8. Details of corner- and inside-angle construction are given below.

## DETAILED INFORMATION ON FORM CONSTRUCTION

*Straight-wall Construction.* (See Figure 17.) There are two accepted methods of constructing a concrete form for a straight wall. If the wall

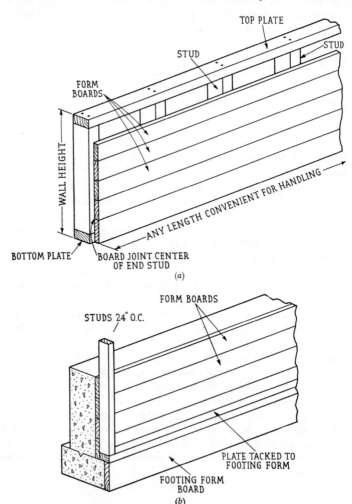

Figure 17. Straight-wall construction: (*a*) panel form, (*b*) continuous form.

is not too long, panels can be constructed by cutting a set of studs the correct length (to wall height) and then nailing 1 by 6 boards to the studs, which are spaced 24 inches o.c. The completed panel is then set to place and properly braced.

If more than one set of panels is required, because of the length of the wall, the boards on the end stud to which the second panel is to be jointed are lapped only halfway across the edge of the stud in order to leave nailing room for the second set of boards. The second panel will then not require an end stud (likewise for succeeding panels).

A second method used to construct a straight concrete-form wall is as follows:

1. Fasten a form plate onto the concrete footing which has already been poured.

2. Mark off stud locations every 24 inches o.c.

3. Cut the studs to correct length (height of the finished wall). The stud layout will show how many to cut.

4. Toenail each stud to place. The studs will stand vertically without difficulty if they have been cut square.

5. Plumb and brace the end studs to exact position.

6. Select a good piece of 1 by 6 stock for the bottom form board; lay against the inside face of the studs and nail, using 6d box nails.

7. Continue placing and nailing boards to place until top of form is reached. The top board may need to be ripped to make the top edge at the exact height of the finished wall. Then brace as described below.

House-foundation forms are often built by using the floor-joist stock. The contractor avoids cutting the joists (so as not to spoil the stock) by extending an outside form several feet beyond the building line.

**Outside-corner Construction.** (See Figure 18.) The simplest way to construct an outside corner of a form is to let one of the forms extend several inches beyond the finished-wall corner. A vertical cleat is nailed to this extension (nails are not driven "home" to facilitate easy removal), and then the boards for the other wall are placed inside this cleat. But few nails are needed on these boards, for the pressure of the concrete will hold them tightly against the edge of the cleat, which should be well nailed.

**Inside-corner Construction.** Inside-corner construction is made by butting the ends of the boards of one of the forms against the ends of the other set of boards. The key to inside-corner construction is to determine which set of boards will be easiest to remove. A vertical cleat is used on the outside face of the boards against which the second set of forms is butted; cleat nails are left for easy withdrawal. The cleat should be

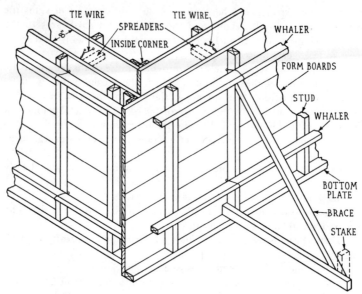

**Figure 18.** Outside- and inside-corner form construction.

well nailed, for it takes the entire pressure of the form during the time the concrete is poured. This type of corner is also illustrated in Figure 18.

*Whaler Construction.* (See Figure 18.) Note that the whaler is placed against the outside edges of the form studs. On a high wall whalers should be placed about every 2 feet o.c.

*Form Ties and Spreaders.* Metal form ties and spreaders are illustrated in Figure 14. The number of metal ties to use is based on the height and length of the wall. The rigidity of the form-bracing structure on the exterior wall is a determining factor. Form ties and spreaders are the mainstay of the *inside* form, which must be held so that it cannot move either out or in.

*Tie Wires.* Tie wires are wrapped around the whalers, crossing from one side of the stud to the opposite side of the opposing stud. A nail inserted where the wires cross enables the carpenter to twist the two wires together to secure maximum strength. This method is illustrated in Figure 14.

*Reinforcing Steel.* Reinforcing steel is hung by wire fastened to the tie wires or metal ties to permit it to become embedded with concrete when the concrete is poured (see Figure 19). Lap joints of reinforcing steel must be as long as 40 times its diameter; that is, a ½-inch reinforcing bar must be horizontally lapped at least 20 inches. Building ordinances are usually explicit on this point. The steel must be laid in the form before the ties or tie wires are placed; otherwise it would not be possible to get it into the form.

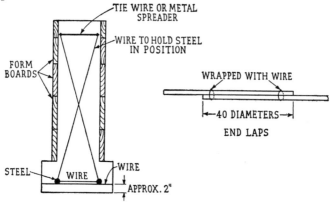

**Figure 19.** Placing reinforcing steel.

*Crawl Hole.* A foundation access hole must be made in every foundation. It should be at least 2′0″ by 2′0″ in size. The contour of the ground may suggest a location at the lowest point to permit a suitable crawl hole. Whenever possible, it should be placed near the rear of the house, preferably in the rear wall.

An opening in a house-foundation wall is formed by placing a finished frame in the form prior to concrete pouring. The frame should be oiled on the inside faces of the jambs to prevent the concrete from sticking and to leave the surface ready for painting. The exterior of the frame must have two beveled cleats, say 2 inches in width, placed vertically, with the smaller face of the cleat against the frame so as to give a dovetail effect when the concrete is poured.

If the finished frame is not available and an opening must be left into which to set the frame, two vertical form boards are placed in the forms at their proper location (see Figure 20). If concrete is to be poured over the opening, a concrete buck is made and the headpiece is made to rest on the sidepieces (see Figure 21). Outside sizes of the buck are ¼ inch larger than outside sizes of finish frame. A beveled cleat is nailed

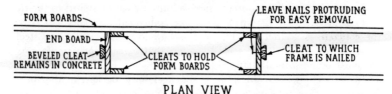

PLAN VIEW

**Figure 20.** "Forming" for an opening in a concrete wall.

to the vertical buck jamb. When the forms are stripped, these cleats will be flush with the concrete wall and ready to have nails driven into them when the frame is set (see Figure 21).

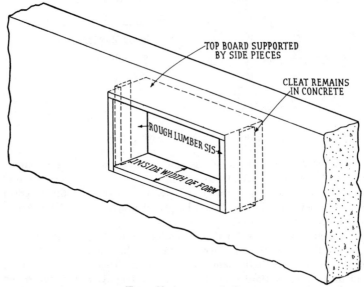

**Figure 21.** A concrete buck.

*Setting Sill Bolts.* Too much care cannot be used in setting sill bolts. The simplest method, but not the best, is to push them into the concrete as soon as the form is filled. The difficulty here is that the bolts may not be set plumb; they may also be too high or too low. Such a condition makes it hard to bolt on the sill, for bolt holes must then be bored at different angles to conform to the bolt angles.

The best method, although not often used by cement contractors, is to make bolt holders from 1- by 2-inch-net pieces of lumber, as shown in

Figure 22. These holders are nailed to the form prior to the pouring of the concrete. In this way one can be quite sure that all bolts will be in the concrete, that they will be in their proper location, and that they will be exactly plumb. The key to this "plumbness" is to be sure that the holes are bored exactly square with the top edge of the holder. Holes are bored approximately 1 inch off center. In order to stagger the bolts, since it is not good practice to place them in a straight line, each holder should be reversed when it is nailed to place. Bolts should be placed not more than 1 inch from each edge of the sill so that the sill will lie flat on the concrete wall. Making the augur bit $\frac{1}{16}$ inch larger than the diameter of the bolt will facilitate the removal of the holder after the concrete is poured.

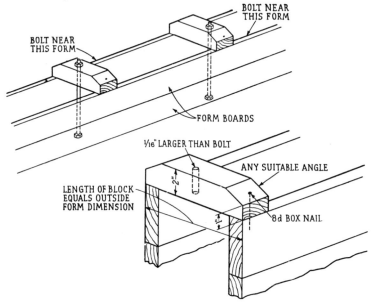

**Figure 22.** Concrete-form mudsill bolt holders.

***Form Construction for Girder-end Support.*** (See Figure 23.) One of the last jobs to do prior to pouring the concrete is to place and fasten blocks in the forms at required locations to provide recesses in the concrete walls on which the ends of the floor-joist girders will rest. These blocks are made slightly larger than the girder, which is usually a 4 by 6. The blocks are also tapered to permit them to be removed easily after the concrete is set.

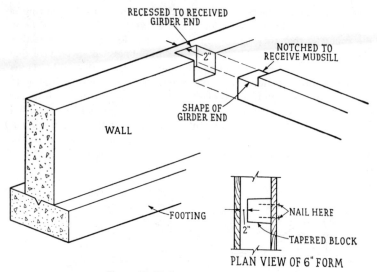

**Figure 23.** Girder-support construction.

The plans must be carefully checked to determine the exact number of blocks required and their exact location. It should be noted that the top of a girder is always flush with the top of the mudsill; hence the depth of the recess will be the same as the depth of the girder minus the thickness of the sill. The girder is notched to allow for the sill. The length of each block is usually made the same as the thickness of the concrete wall less 2 inches.

The blocks are nailed to the inside form to make them secure during the concrete pouring. Since they can be fastened only at one end, it is advisable to do this job just prior to pouring concrete.

*Piers.* Concrete piers are required to support the underpinning structure, which in turn supports the floor joists (see Figure 24).

If a house has no basement, ordinance requirements usually call for piers to be spaced a maximum of 5'0" o.c. when a 4 by 4 girder is used and a maximum of 7'0" o.c. when a 4 by 6 girder is used.

A slab floor requires no piers (see Figure 26).

The size of a pier, that is, the dimensions of the pier at the bottom, varies with the load. In some localities where partition loads occur 16 by 16 inches is standard. No definite rule or dimension can be given here, but building ordinances are very specific in this regard as they have been carefully engineered to fit the local conditions. Obviously the condition of the soil is another important factor. A soft soil would require a larger pier base than would a hard soil.

The height of a pier is usually 12 inches. The top dimension will be, on the average, 6 by 6 or 8 by 8 inches. The smaller dimension at the top results in a tapered pier.

An average 1,000-square-foot house may require as many as 18 to 20 piers (assuming that there is no basement). The required depth of the

Figure 24. Concrete piers.

bottom face of a pier below undisturbed ground is governed by the local building code.

For houses with basements, where heavier girders are used in order to eliminate posts, piers are made larger and are spaced farther apart.

Piers are often precast by means of pier forms and allowed to harden. They can then be handled without difficulty and set to place.

Sometimes two or three pier forms are made and, after the center line of the piers has been located, are placed in position and filled with concrete, which is purposely made dry for quick setting. The pier forms are filled during the pouring of the footings and walls, allowed to set for a brief time, carefully lifted (the taper makes this an easy operation, provided that the concrete is not too "runny"), and then reset for a refill.

To locate the position of each pier, proceed as follows:

1. Read the blueprint and, by means of a steel tape, locate the center

line of a given row of piers by driving a nail into the top edge of one of the inside forms.

2. Repeat this process on the form opposite.

3. Continue this procedure until the centers of all rows of piers have been located.

4. Next stretch building lines from form to form to represent the center of each row of piers.

5. Determine from the blueprint the center of the first pier from the wall form. Measure this distance on the building twine (which must be taut); hold a plumb bob at this point; then carefully lower the plumb bob, avoiding a swinging motion, until the ground is touched. At this point drive a small stake which represents the exact center of the base of the pier.

6. Continue this process until all piers are located.

7. Carefully dig the earth at these stake locations until solid earth is reached.

**Note:** Where there is very loose soil, first take approximate measurements, determine approximate pier locations, and remove the dirt until good solid ground is reached; then accurately locate each pier center as described above.

There is no need to get the top of each pier exactly level with the other piers; any differences in height are adjusted by cutting each post an exact length to make the top of all posts level in readiness for the girders to be placed and nailed into position.

Pier forms are constructed by making two sets of pier panels the exact size of the pier dimensions, say 16 inches long at the bottom, 8 inches long at the top, and 12 inches wide. The second set has the same dimensions *plus* twice the thickness of the form material; that is, for a 16-inch base the form is cut 17½ inches long at the bottom and 9½ inches long at the top. These pieces are then nailed together to form the tapered-pier form. If narrow boards are used, they will need to be cleated on the outside to hold them together (see Figure 25).

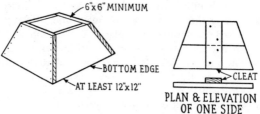

**Figure 25.** Concrete-pier forms.

**Note:** A tapered pier is actually in the form of a hopper. Hoppers—to be accurate, with all joints fitting—require a geometric layout, which can be made by means of the steel square. No instruction is given here on the steel-square method since the method already described is sufficiently accurate to serve as a practical layout procedure for concrete piers.

## SLAB-FLOOR OR BASEMENT FORM CONSTRUCTION

There is no actual form-construction work for the carpenter to do when preparations are made to pour a concrete floor slab. Strips of lumber (called *screeds*) are placed by the concrete man during the process of pouring, spreading, and troweling the concrete floor. The screeds act as guides on which a straightedge is pushed backward and forward to attain the correct thickness and pitch of the floor. They are removed after the initial set of the concrete materials, and the space is filled with the concrete mix.

It is very important to make sure a watertight joint is obtained at the junction of a concrete floor and a concrete wall (see Figure 26). Three beveled pieces of lumber, actually siding boards, are oiled or soaped thoroughly and then placed next to the wall, as shown in the illustration. The bevel of the lumber permits the easy removal of the three strips after the concrete is set. The space occupied by the strips is then filled with hot tar, which makes the joint absolutely watertight by preventing the seepage of water from beneath the floor.

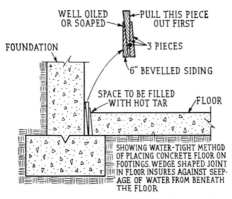

**Figure 26.** Form construction for floor-and-wall watertight joint.

# CHAPTER SIX

# Framing

The completed frame of a new house may appear to be the result of merely repeating a large number of simple carpentry operations, such as cutting and nailing. Actually, however, it is the result of numerous jobs, many of which required a knowledge of blueprint reading, layout, selection of the correct kind and size of lumber, cutting lumber, and nailing or assembling the various parts into a complete whole.

To simplify the explanation of the various construction processes, the framing procedure has been divided into five units, or divisions, as follows: underpinning, floors, walls, ceiling, and roof.

Emphasis must again be given to the value and requirements of a building code. The carpenter in charge of a framing-construction job, whether an expert, a trained craftsman, or a homeowner doing his own building, must have an intelligent understanding of the code if he expects to receive inspection clearance without difficulty.

No building permit is issued unless a drawing is presented to the building inspector. A carefully prepared drawing will be based on the exact requirements of the local building code. However, some construction methods which are shown in detail in the building ordinance may not be "spelled out" on the drawing. It is up to the building inspector to be sure that these requirements are also followed.

The various jobs described in this chapter are based on code requirements covering sizes of material, methods of framing, size and number of nails to use for each joint, spacing of framing members, and bracing minimums.

Ordinances will vary between cities and between states. Eastern construction is different from Western construction. It is therefore imperative that a builder secure copies of the building ordinance that applies to the type of work he expects to do. Local requirements have a priority over any detailed drawing given in this chapter. The construction procedures illustrated in the following pages are basically sound, but no one method can be shown that will fulfill every specific requirement in the country.

## U N D E R P I N N I N G   U N I T

The term *underpinning* refers to building-parts construction on a concrete foundation to form a support for the floor joists (see Figure 1). Underpinning includes mudsill (in the early days of house building the sill was actually placed on the ground), pier blocks, girders, and posts. Each of these items is described below, but it should be fairly obvious that no description of a particular situation can cover every other situation. For the man building a simple mountain cabin the underpinning requirements may be quite simple as compared with the jobs involved in building a pretentious home on a hillside. Fortunately, there are a few basic principles which can be applied to any type of foundation work. Hence the following explanation should be considered typical of the average underpinning job.

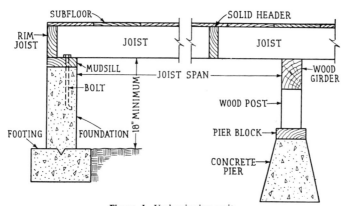

**Figure 1.** Underpinning unit.

*Mudsills.* As described in Chapter 5, The Foundation, mudsills (called sills for short) are usually bolted to the foundation. Occasionally a builder will spike the sill to the concrete wall, driving the nails while the concrete is still "green." This is not considered good practice, and in most communities that operate under a building code it is not permitted. It is an established fact that in areas where there are occasional earthquakes the bolt method provides the only means of preventing houses from being shaken off their foundations.

The best sill material is either redwood or cypress. It should be S1E to give it a uniform thickness, and some codes require that the underside of the sill be creosoted to prevent later damage by termites. Building codes are quite definite about the requirements on this point.

Sill material should be ordered in lengths selected to avoid too many joints and too much waste. Hence the foundation plan should be studied to determine the best lengths to order for the various wall dimensions. Most sill lumber does not exceed 20 feet in length; hence a wall 46 feet long would require, say, two 14-foot lengths and one 18-foot length of sill stock or, say, two 16-foot lengths and one 14-foot length. Any combination will do that is economical, avoiding loss due to waste ends.

Sill material should be reasonably straight; hence if a crooked piece of sill stock is delivered to the job, it should be rejected or, if possible, cut into short pieces, which may be required for short walls.

A butt joint is used at all corners and inside angles (see Figure 2). The direction of the joint is not important as the end of one of the two pieces which form a right angle will be bolted if the bolts were carefully located. The end of the other piece is toenailed to the first one, thereby holding both pieces securely in place.

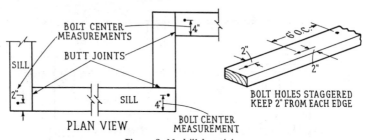

**Figure 2.** Mudsill butt joint.

To locate the position of the bolt holes in the sill so as to permit the sill to pass over the bolts and lie flat on the wall, proceed as follows:

1. Lay a piece of sill stock on the wall *against* the bolts with one end placed in exact position, usually flush with the outside face of the concrete wall. It may be necessary to provide temporary supports to hold the sill in place since the bolts prevent full flat contact with the wall.

2. Using a try square, carefully mark on the smooth side of the sill the center of each bolt (nuts should have been removed for accuracy in marking).

3. Measure from the face of the concrete wall the exact distance to the center of each bolt (see Figure 2) and mark carefully on each line made on the sill stock. These measurements will be different for each bolt if the bolts have been "staggered" as described in the previous chapter under Setting Sill Bolts, page 71. This procedure assumes that the foundation

wall is perfectly straight—as it should be if the forms were accurately constructed. The wall should be sighted for straightness; if it is not straight, then stretch building twine along the wall, fastening it to the batter boards. It is then an easy matter to measure from the line to the center of each bolt and mark on the sill the location of the center of each bolt hole.

All bolts should be in an exact vertical position, as they will be if the bolt-holder method is used. If the bolts are not exactly plumb, they should be hammered carefully until they are plumb. Do not pound on the bolt threads as this will ruin them.

Note: Occasionally it is advisable to keep the outside edge of the sill at least 1 inch "in" from the face of the wall. This is to allow for the thickness of sheathing stock on the outside face of the framed walls (see Figure 3).

4. Holes are now bored where marked. It is best to use a bit $\frac{1}{16}$ inch larger than the bolt diameter to allow for slight discrepancies in layout.

5. The sill is now ready to be placed on top of the bolts, and it should take but little effort to drop it into place.

6. Continue with sill piece 2; then 3, etc., until all pieces have been cut to exact length and bored ready for placement.

7. The sill stock should now be creosoted, if required. Gloves must be worn during this operation and great care used to keep the creosote from being splashed into the eyes or on the bare skin, as it can cause very painful burns.

8. Next, before the sill is set, a cement grout is prepared and placed on the wall.

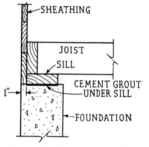

Figure 3. Illustrating mudsill offset.

This grout allows for any unevenness in the top of the concrete wall. Grout is composed of equal proportions of screened sand and cement, which are mixed together until the material has a fairly good "running" consistency.

9. Wet the top of the wall thoroughly; place the grout in the center of the wall for a distance equal to the length of one piece of sill stock; put the sill stock on top of the bolts and gradually push into place. Some grout will squeeze out, but enough will remain to give a good solid base to the sill. Carefully pound all high points on the sill, checking with a straightedge until a perfect flat bed is obtained (see Figure 4). Then put on the bolt washers and nuts and tighten carefully with a wrench. Finally, remove excess grout with a small pointing trowel.

10. Repeat this process until the entire sill is bedded and bolted. All

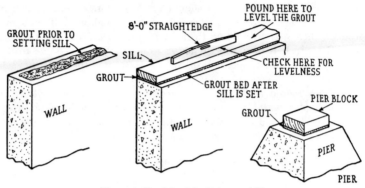

**Figure 4.** "Straightedging" the mudsill.

butt joints should be carefully toenailed, with four 8d common nails used for each joint. The straightedge and level should be used constantly during the grouting process to obtain an accurate and level sill job. Sufficient time spent to secure an accurate job will save much time later, for the joists can then be set on the sill with the full knowledge that they will make a true and level subfloor.

11. The sill will need to be cut where it passes over the recesses made in the concrete wall to receive the girder ends. The notches are best cut after the sill is grouted and bolted, so that an accurate girder-layout job can then be done.

**Pier Blocks.** Pier blocks are made from waste ends of sill stock and the poorest pieces of mudsill, usually the crooked ones (see Figure 4).

The number of blocks required is the same as the number of piers; the sizes are as shown on the blueprint or as required by the code. They are usually 6 by 6 or 8 by 8, depending on the dimensions of the pier on which they are to be placed.

Each block is grouted to the pier in a manner similar to the one used for the sill: wet the pier, place a small amount of grout on it, and press the block into the grout, testing it for levelness in both directions and tapping it lightly until this effect is obtained. Continue this operation until all pier blocks are set.

**Girders and Posts.** Girders vary in size (thickness and width) according to the span distance from bearing to bearing (see Figure 1). Building codes are very specific about girder requirements; they give dimensions on the basis of the load to be carried as well as the allowable maximum span.

For a building without a basement, piers can be placed almost anywhere and as close together as necessary. In houses with basements, girders are usually supported by the basement walls. If the span is too great to permit a girder of reasonable size, an intermediate post or Lally column (steel post) is used (see Figure 5). This post is supported by a pier of suitable size to carry the load.

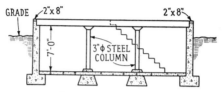

Figure 5. A Lally column.

Douglas fir, eastern hemlock, western hemlock, and western larch are used for girder material. Girders and beams can be either of one piece, such as a 4 by 6 S1E (if all are to be the same width) or built up of 2-inch stock to the required dimension, such as 6 by 12 or 6 by 14. For extra strength it is best to use carriage bolts when constructing a built-up beam. Bolts should be placed not more than 1½ inches from each edge and approximately 24 inches o.c. horizontally (see Figure 6).

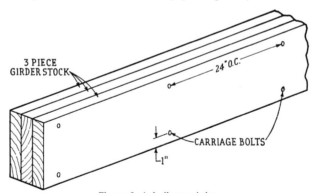

Figure 6. A built-up girder.

To cut and set girders to support the floor unit of a house, proceed as follows (assuming that the sill stock is already bolted to the foundation walls, all piers and pier blocks are in place, and foundation-wall recesses are cut and ready to receive the girder ends):

1. Place girder stock inside the foundation in the approximate location of each girder.

2. Stretch building twine from wall to wall directly over one line of piers. Be sure that it is stretched tight to make a taut line. Even then ¼ inch must be allowed for line sag when measurements are taken from the line to the pier blocks.

3. Select a pier on which the end of a girder is to rest and measure from the line to the pier block; allow ¼ inch for sag and then *subtract* the width of the girder. The result equals the length of the girder post which is to rest on that particular pier (see Figure 7).

4. Cut a post this length. Posts are usually 4 by 4, except in basements, where they may be larger because of their increased length. Posts should be carefully cut so as to make a true and square cut.

5. Toenail the post to the pier block, using four 8d common or 8d box nails.

6. Cut the girder to length so that the joint will center on the post, thus leaving the other half of the post to receive the end of the next piece of girder, which is to be butted and joined to the first one (see Figure 7). The wall end must be cut to fit around the mudsill in order to make the top of the girder flush with the face of the mudsill.

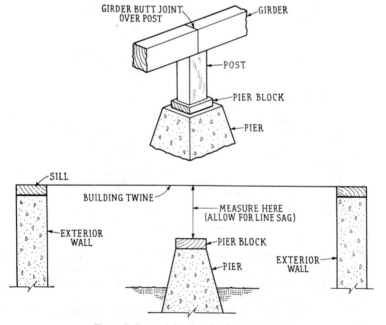

**Figure 7.** Determining length of a girder post.

7. Place the girder in position—one end on the wall and the opposite end on the post—and nail, using two 8d common nails at each end. Be sure the "crown" or rounded side of the girder is "up."

**Note:** A girder that has a very high crown should not be used unless it is partially cut over one of the posts on which it is to rest. Cutting the girder will relieve the crown and thus straighten the piece of lumber.

If the post is fairly high, it may be necessary to brace it temporarily to keep it immovable until all girders are in place.

8. Continue with the other girders until all are in position.

9. Next measure from the remaining pier blocks to the underside of the girders cutting the various posts to the different lengths measured, and nailing the posts to the pier blocks and girders. The ends of the girders that butt together should be nailed with two or more 16d box nails.

10. The finished job should now be sighted as a final checkup. A long straightedge, preferably a selected floor joist, is very useful for making a final check of the girder alignment. Since the accuracy of this job greatly simplifies the next job, that of cutting and placing the floor joists, any discrepancies should be corrected. At this stage of underpinning construction no wedges should be used under pier blocks which may have been raised because of the crown in a girder. *After* the floor joists are on, check each post for immovability, which indicates that it is carrying a load, and carefully wedge the posts that are loose from the pier.

## FLOOR UNIT

The floor unit, as described below, includes joists, header joists, bridging, framed openings for a fireplace or a stairway, and the subfloor. This "unit," which is constructed of many pieces of lumber, is supported by the underpinning and, in turn, carries all walls and partitions.

A two-story house has two floor units. The upper one is essentially a duplicate of the lower one; the basic difference is merely in the width of the joist stock, which is determined by the dimensions of the room which the joists span.

Second-story joists will vary in size, such as 2 by 8, 2 by 10, or 2 by 12; for very wide spans the joist stock may be 2 by 16. It is easy to determine what size of joist material to use, for building ordinances include tables giving various joist sizes for specified span dimensions. Properly prepared plans also indicate approved sizes.

Construction procedures for a second-story floor unit, although the same as for the first floor, are more difficult, as the work must be done from a scaffold or "high horse" or ladder. The following description is limited to the one-story house but includes the full-basement type of building. Joist setting over a basement is typical of second-floor-unit joist cutting, fitting, and nailing.

Floor joists usually "run" the shortest dimension of the building. This practice is based on the engineering rule that "the shorter the span, the more rigid the floor."

***Floor-joist Layout.*** Floor joists for the first story of a house can be spaced as desired up to the maximum allowed by the ordinance. It is customary, however, to space the joists 16 inches o.c., which, incidentally, must be the spacing of second-floor joists, as they also serve as ceiling joists. This 16-inch spacing is based on the sizes of plasterboard, metal lath, or wood lath. Plasterboard is made 16 inches wide and 48 inches long; metal lath is 24 inches wide and 96 inches long; wood lath is 48 inches long. Sixteen-inch o.c. spacing exactly conforms to these lath lengths, which, as noted, are made in some multiple of 16.

Joist layout can be started from any wall. If a building is L-shaped, however, or if there are insets in an exterior wall so that it is not continuous, always start the layout from the longest unbroken wall.

Ordinance requirements state that joists must be doubled under every cross partition. (A cross partition runs parallel to the joists.) Joists are doubled and well spiked together or separated by solid blocking, not more than 4 feet on center, to permit the passage of pipes (see Figure 8).

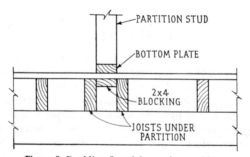

**Figure 8.** Doubling floor joists under partitions.

To simplify joist layout the expert carpenter will make a sample layout on a piece of 1 by 4 of suitable length for handling, say 12 or 16 feet (see Figure 9). This pattern, which is called a joist-layout rod, is then placed on the sill and the joists marked; the rod is shifted to a second position and the operation repeated. Then the rod is moved to the opposite wall and the same process repeated. This procedure ensures that the joists will be exactly parallel to the outside walls. The carpenter must take care, however, to see that the rod is *never reversed* end for end while he is using it.

The steel square is indispensable when laying out the locations of the joists, providing that it has a 16-inch tongue (some squares are made with 18-inch tongues).

In the following description of the steps to follow when laying out the location of the floor joists on the sill, the use of the rod will be included since the procedure is basically the same whether the joist locations are marked on a rod and then transferred to the sill or the layout job is done directly on the sill.

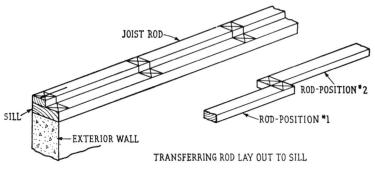

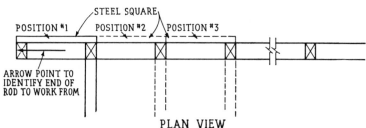

**Figure 9.** Using floor-joist layout rod.

1. Select a straight-grained piece of 1 by 3 stock 16 feet long. It should be S1S. Lay it on a pair of sawhorses, preferably on a plank to prevent sagging.

2. Work from the left end of the piece toward the right. The starting end should be tested and, if necessary, cut to be sure it is square. Use the body of the steel square and mark for the end joist.

3. Turn the square over and place the end of the tongue of the square even with the end of the 1 by 3; then mark both sides of the

square blade (see position 1, Figure 10). These marks represent the second joist.

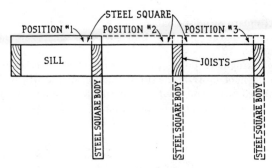

**Figure 10.** Using steel square for joist layout.

4. Move the square to mark for the third joist, always keeping the end of the tongue even with the joist mark nearest the beginning end of the rod.

5. Continue in this way until the end of the rod is reached.

6. Make an arrow point (see Figure 9) to identify the end of the rod from which the layout work was started. This mark is *very important* because the rod, whenever used, must be placed with the arrow point at the left end. If the rod were reversed during the layout, the sill layout would be totally incorrect on opposite walls.

7. The rod is now ready for use on the sill. Place the marked end of the rod even with the outside edge of the sill and transfer all marks to the sill.

8. Move the rod so that the left end coincides exactly with the last joist marks (on right end of the rod) and repeat the process, continuing until the end of the building is reached (see Figure 9). A set of joist marks must also be made at this end of the building to represent the end joist.

9. Repeat this entire process on the sill opposite the first one laid out. It is again important to stress the fact that the *joist layout must be started from the same outside wall*. In this way one can be definitely sure that all joist marks will be parallel to the outside walls.

10. Outside walls which are not in a continuous straight line must be laid out so that the 16-inch o.c. spacing is not changed. To illustrate, if the last joist comes 10 inches from an inside angle, then the next joist will be 6 inches in from the end of that part of the building. In other words extra joists are required whenever a break in a wall occurs (see Figure 11).

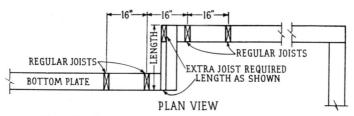

PLAN VIEW

**Figure 11.** Joist layout for outside walls that are not continuous.

**Note:** The layout as described above provides for two marks, spaced 2 inches apart, for each joist. Actually joist stock measures only 1⅝ inches in thickness. Many carpenters make one mark and an *X*, as illustrated in Figure 12. Joists are then placed in position by means of the single mark. If *two* marks are made, then the joist must be centered between the two. For a beginner, two marks are recommended to avoid the error of placing the joist on the wrong side of a single mark, as a result of either forgetting to mark the *X* or overlooking it.

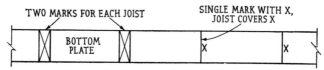

**Figure 12.** Alternate methods of joist layout.

11. Now check the blueprint floor plan and, using a steel tape, mark the centers of all cross partitions on the sills. Since double joists are required under each cross partition, these additional joists must be located and marked (see Figure 8). Sometimes a regular joist will serve as one of the "doubling" joists.

12. Girders are rarely marked for joist locations since the joists can be located with sufficient accuracy by sighting across the top edge of the joist from sill to sill as the joists are placed and nailed. Narrow buildings are constructed with a single length of joist stock. At one time it was common practice to use joists for first-floor construction from 28 to 32 feet in length. In more recent years it has become common practice to use shorter lengths and lap and spike them together over a girder (see Figure 13).

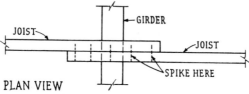

PLAN VIEW

**Figure 13.** Floor-joist lapped joints.

*Floor Openings.* The joist layout is now complete except for locating floor openings for a fireplace or stairway (see Figure 14). The two joists which form the sides of a framed floor opening are accurately located from the blueprint. These two joists must be doubled *after* the opening is framed in.

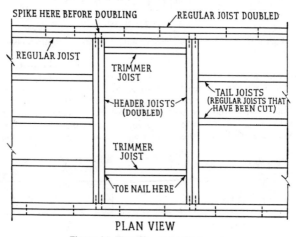

PLAN VIEW

**Figure 14.** Floor-opening joist layout.

When locating a floor opening, work from its center and measure in opposite directions to locate the inside edges. This procedure is necessary to make sure a fireplace, for instance, is located exactly in the center of the room (assuming that this is where the plan shows it to be).

Regular joists must be cut to provide the floor-opening space. To provide a practical means of supporting the ends of these joists (called tail joists), header joists are framed in between the two regular joists forming the ends of the opening. The header joists are also doubled to give added strength.

After this framing work is completed, two trimmer joists are framed and toenailed in between the two header joists to make the opening the correct length. Figure 14 gives a plan view of a framed opening.

To lay out a stairway opening requires a knowledge of the stair layout, since the length of the opening, called a wellhole, is determined by the number of treads (steps) and the pitch or slope of the stairway. The opening must also give sufficient headroom so that the user can walk in an upright position while ascending or descending the stairs.

The stair-layout problem is sufficiently complicated to require much more discussion than is possible in this book. The work is usually done by a specialist known as a stairbuilder. Simple open stairs, however, such as are required for a basement, are not difficult to construct provided that one has a basic knowledge of the principles of stair layout.*

*Box Sills.* (See Figure 15.) A box sill is constructed by placing a continuous joist flush with the outside edge of the sill stock on the two walls which are to support the ends of the floor joists. The continuous joists which rest on the sill and the end joists are sometimes called rim joists since they actually form a rim around the building. The floor joists are butted and nailed to the rim joists to prevent them from twisting.

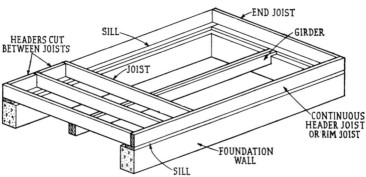

Figure 15. A box sill (alternate methods).

If continuous rim joists are not used, the alternate method is to cut headers *between* the floor joists.

Building ordinances are quite specific on the topic of header requirements on the exterior walls. Solid blocking, which is cut *between* the joists, permits the ends of the joists to have a full bearing on the sill,

---

* See J. Douglas Wilson and S. O. Werner, "Simplified Stair Layout," Delmar Publishers, Albany, N.Y., 1946.

whereas if the rim-joist method (see Figure 15) is used, the bearing will be 2 inches less. To illustrate, if a 2 by 6 sill is used, joists will have a full 6-inch bearing, because they extend to the outside edge of the wall and the solid headers are cut in between each joist. If a rim joist is used, the joists have a 4-inch bearing. In many ordinances the full-bearing method is required.

*Setting Floor Joists.* Assuming that the entire joist layout has been completed, which includes the extra joists required under cross partitions and the key joists which form the two ends of the floor opening, proceed as follows to set the floor joists.

1. Select straight pieces of joist stock for header or rim joists and cut and nail to place. This includes the two end joists. When this job is completed, the floor area should be completely enclosed. Use 8d common nails and toenail from the outside and lower edge of the joist into the mudsill, spacing the nails about every 24 inches. Spike corner joints with three 16d common or box nails.

If no rim joist is used, check the outside end of each floor joist for squareness. Then lay the joist on the sill with the end exactly flush with the outside edge of the sill. Drive three 8d box nails into the joist, one through the end and one on each face of the joist.

Cut solid headers to fit tightly in each of the joist spaces but leave out certain pieces to provide foundation ventilation (see page 103 for further information on foundation vents).

2. Using the steel square, make vertical location lines for each joist on the inside face of the rim joists. This is easy to do, for it merely means continuing the horizontal lines on the sill to the inside face of the rim joist.

3. Lay the joists in a flat position on the sill and girders with the crown edges facing the same direction. This is done by sighting along the edge of each joist and determining which is the rounded, or crown, edge (see Figure 16).

**Note:** If all joists are laid flat with the crown edges all one way, there is no need for further sighting when turning them on edge preparatory to nailing. Otherwise much time is spent in checking the crown of each joist at the time of nailing.

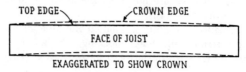

EXAGGERATED TO SHOW CROWN

**Figure 16.** Placing floor joists crown side up.

4. Cut off excess joist material that extends beyond the girders, leaving enough, however, to make a good lap joint, say 12 inches (see Figure 13). If the joists are long enough to go from rim joist to rim joist, then carefully cut each joist to the exact length to drop between the rim joists. If a joist is cut too long, it will crowd the rim joist out of line, and the result will be a crooked outside wall.

Toenail the joists to the sill, using two 8d common nails per joist. Also drive two 16d common or box nails through the rim joist into the end of each joist and nail the joists to the girders, using two 8d common nails for each joist on each girder that it rests on. Spike lap joints together with 16d box nails. The length of the lap determines how many nails to use; there should be at least three. The nails driven into the girder will hold the joists together at that point.

Any waste piece that is equal in length to the spacing between the joists (usually 14 inches) is worth cutting off since it can be used for solid header stock (see Figure 18).

5. Now frame the floor openings as outlined above.

6. As a final check, select a good straightedge and test the surface of the joists for evenness, adjusting high joists by chiseling under them where they rest on the sill or girder or, reversing the procedure, raising any joist that may be low (see Figure 17). If the work of joist setting has been carefully done, little adjustment should be needed, for "extra-crown" joists will have been corrected and very crooked joist stock eliminated or used for short joists. A few minutes with a good straightedge, however, will quickly show if and where minor adjustments are needed.

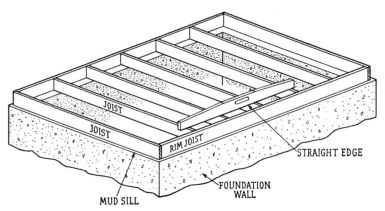

**Figure 17.** Testing floor-joist surface with straightedge.

*Bridging.* *Bridging* is the term applied to joist material that is nailed between the joists where the span is wide and the joists need stiffening to make them more rigid and to enable them to act as a unit.

There are two types of bridging, solid and herringbone. Solid bridging is made from material the same size as the floor-joist stock and is required over all main bearings, as well as halfway between these bearings if the span is more than 8 feet. A center line is established on the joists that are to be bridged by measuring equal distances from the joist bearing or, to state it more simply, half the span. The solid headers are then alternately nailed on each side of the line, thus permitting the driving of two 16d box nails into each end of each piece of bridging (see Figure 18). Solid headers must be kept flush with the top edges of the joists.

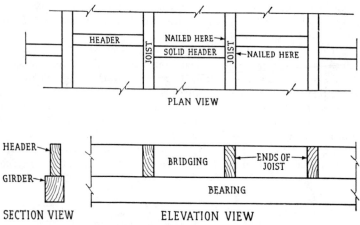

**Figure 18.** Solid bridging (or header).

*Herringbone bridging,* a trade term, is illustrated in Figure 19. In this case, 2 by 3 or 2 by 4 stock is cut to the correct bevel and length so that when the bridging is in place each pair that meets at the top edge of a joist acts as a small "truss," preventing the joist from sagging (see *A-A,* Figure 20). Herringbone bridging is most often used with wide joist stock, such as 2 by 12 or 2 by 16. Wide spans may require two or three rows of bridging; the local ordinance must be checked on this point.

It is important to note that the lower end of each piece of herringbone bridging is *not* nailed until after the subfloor stock is laid (nails are shown in the illustration). This practice permits uneven joists, those with crown or sag, to level up a little when stress is placed on them during the process of nailing on the subfloor. While it is not easy to do when

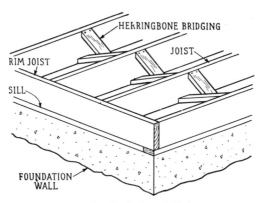

**Figure 19.** Herringbone bridging.

the space is crowded, the carpenter must crawl under the house to complete the bridging nailing. In a two-story house a stepladder will be needed for the second-floor joist bridging. Two 8d box nails are required for each end of a piece of bridging. Nails are driven at right angles to the cut.

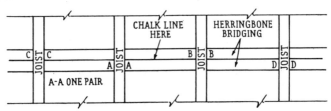

**Figure 20.** Placing herringbone bridging.

The layout for a piece of herringbone bridging is illustrated in Figure 21. Proceed as follows:

Bridging can be laid off by using the carpenter's steel square (see Figure 21).

The bridging-cut problem is simply one of rise and run. As shown on the drawing, the rise of the bridging is not the actual width of the joist stock but is at least 2 inches less than that width.

The run is equal to the width of the spacing between the joists.

For a floor using a 2 by 10 joist, set 16 inches o.c., the run would be 14 inches while the rise would be 7½ inches.

Since the exact difference between distances *B* and *C* in the drawing is not known until one piece of the bridging is cut and the length of the plumb cut can be measured, the simplest way is to assume that the width

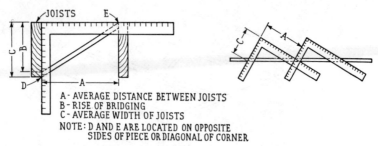

A- AVERAGE DISTANCE BETWEEN JOISTS
B- RISE OF BRIDGING
C- AVERAGE WIDTH OF JOISTS
NOTE: D AND E ARE LOCATED ON OPPOSITE
        SIDES OF PIECE OR DIAGONAL OF CORNER

**Figure 21.** Layout for herringbone bridging.

of the joist is the rise of the bridging. Then, using this rise and the spacing as the run, lay off the pattern, keeping the rise and run figures on the *opposite* side of the bridging stock. This will automatically take care of the difference between *B* and *C*.

Shift the square to its second position and mark another plumb line. The layout of the pattern is then complete.

To use the pattern, proceed as follows:

1. Select the bridging material. Often short ends of suitable stock are available. The bridging job, in fact, provides an excellent opportunity to "clean up" the waste ends of material which have accumulated (a never-ending procedure).

2. Use the bridging pattern, and mark and cut the desired number of pieces. The number of joist spaces times the number of rows of bridging times 2 equals the required number. There will be some spaces that are not the same as the regular spacing because of joist doubling. For these narrower spaces the bridging pattern will not fit; allowances must therefore be made and special pieces cut.

3. Drive two nails in each end of each piece at right angles to the cut in readiness for nailing to place. This permits placing a piece of bridging in place and then driving the nails; otherwise the carpenter would have to hold the piece and try to start the nails all in one operation.

4. Strike a chalk line to represent the center of each row of bridging.

5. Nail the top ends of the bridging to the line, alternating pairs on each side of the line (see Figure 20).

6. As noted above, nail the lower ends after the subfloor is laid.

*Foundation Vents.* If foundation vents, as described in Chapter 5, pages 71–72, have been placed in the concrete foundation no further joist work is necessary. If the vents are to be placed in the joist structure, proceed as follows:

1. Check the plans or the building code (they should coincide) to find the number of vents required.*

2. Locate, on the rim joists, the vent openings. The vents for the exterior walls on which the ends of the floor joists rest are located by selecting certain joist spaces. The vent openings for the exterior walls on which the end, or last, joist rests must be laid out and marked. The size of the opening is adjusted to fit vent dimensions.

3. Cut out the vents where marked.

4. Fasten the metal vents to place.

**Note:** For the walls on which the joist ends rest, it will be necessary to frame in a small piece of 1 by 2 stock between the joists to provide nailing for the diagonal subflooring where it passes over the vent openings (see Figure 22).

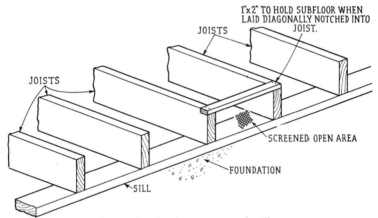

**Figure 22.** Floor framing over screened wall vent.

*Exterior-doorsill Framing.* All exterior doorframes are constructed with a doorsill (see Figure 41, Chapter 7, page 216), usually hardwood, to cover the joist framing, to furnish a sloping area for the drainage of rain water, and to provide a finished appearance at the bottom of the door opening. Space for this sill must be planned for and framed into the floor joists.

The length of the space equals the *outside* width dimension of the doorframe. In some instances an extra joist may be required to form the recess end. The width of the space is the same as the thickness of the framed wall plus the thickness of the interior plaster. To illustrate, a

---

* For example, the Division of Building and Safety of Los Angeles County requires 2 square feet of opening for each 25 linear feet of wall. If 2 by 6 joists are used, a screened vent of maximum size (joist width times joist spacing, or 5½ inches times 14 inches) is required approximately every 6 feet.

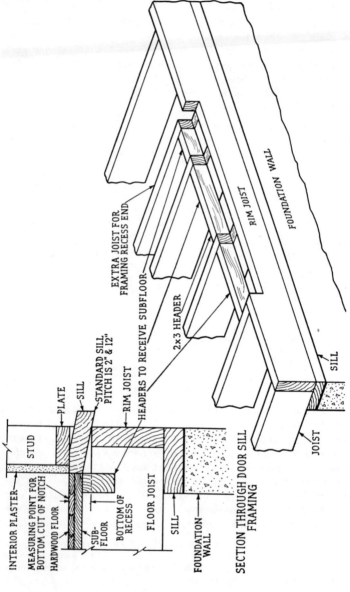

INTERIOR PLASTER

STUD

PLATE

SILL

STANDARD SILL,
PITCH IS 2" & 12"

MEASURING POINT FOR
BOTTOM CUT OF NOTCH

HARDWOOD FLOOR

RIM JOIST

HEADERS TO RECEIVE SUBFLOOR

SUB-
FLOOR

BOTTOM OF
RECESS

FLOOR JOIST

SILL

FOUNDATION
WALL

SECTION THROUGH DOOR SILL
FRAMING

EXTRA JOIST FOR
FRAMING RECESS END

2x3 HEADER

RIM JOIST

FOUNDATION WALL

SILL

JOIST

**Figure 23.** Floor-joist framing for exterior doorsill.

wall using a 3⅝-inch stud plus ¾ inch for interior plaster requires a sill space of 4⅜ inches from the outside face of the wall. The depth of the recess equals the thickness of the sill plus the pitch of the sill. This depth measurement is taken *from the finish-floor surface* (see Figure 23).

To lay out the recess proceed as follows:

1. Determine the exact location of the doorframe from the blueprint. It is best to work from the center of the frame and measure an equal distance in each direction from this center mark.

2. On the top edge of the floor joists into which the recess is to be framed, mark a cutting line ¾ inch in from the inside line of the wall plate.

3. Saw the joists at this point, to a depth equal to the sill thickness (plus pitch). Be sure, when measuring the depth of the recess, to allow for the combined thickness of the subfloor and the finish hardwood floor as the face of the sill must be flush with the hardwood floor surface. After marking, chisel out the wood from the saw cut to the end of the joist.

4. Cut in headers between the floor joists to provide nailing for the subflooring where it passes over the recess.

*Laying Subfloor.* Subfloor is the term applied to the rough floor which serves as a base or nailing surface for a hardwood or other type of finish floor (hence the prefix *sub,* meaning "under").

The lumber most commonly used for subflooring is 1 by 6 S1S1E. Using sized stock (of the same width) results in a better job as the boards will then lie close together. The lumber may be one of several kinds, such as white fir, eastern hemlock, Douglas fir, or western larch. It is customary to use the subflooring stock ordered for the job as concrete-form material, recovering as much as possible for later use as flooring. If form lumber is used as flooring, it should be carefully cleaned for two reasons: (1) concrete material left sticking to the side of a board placed on the joist will make an uneven floor surface; (2) concrete material adhering to a board will quickly dull the crosscut saw or power saw.

Which side of the flooring stock should be placed up, the smooth or the rough face, is a debatable question. Some carpenters state that the smooth side will give a surface on which the hardwood will lie flat. This fact is no doubt true. It is also a proved fact, however, that subfloor stock which is *laid rough side up* will *not warp* during the framing process, when the floor is exposed to the sun. Hence a surface composed of 1 by 6 boards which are not warped makes a much more satisfactory base for a hardwood floor than would result if the smooth side were placed up and the boards became warped.

Subflooring is laid in one of two ways: (1) at a right (90-degree) angle to the joists or at a 45-degree angle to the joists. Either method is satisfactory, but the direction of the hardwood floor must be considered since it is not good practice to lay the hardwood floor in the same direction as the subfloor.

The direction of the finish flooring affects the artistic appearance of a room. No rule can be given regarding how a hardwood floor should be laid, for there are two generally accepted principles, both of which obviously cannot be followed at the same time: (1) the floor should always run the long dimension of the room; (2) the floor should run at right angles to the front door; as you enter a room, you would walk in the length direction of the flooring.

The 45-degree-angle method is recommended for laying the subfloor since it permits laying the hardwood floor in either direction.

**Note:** It might appear that a floor laid diagonally would provide a stronger job because of its bracing effect. This is true, but a floor unit properly bolted to a concrete foundation obviously is so rigid that no additional bracing is necessary.

Subfloor is laid at a 45-degree angle, as stated, to provide the best finished appearance for the hardwood floor, not to strengthen the building.

In a two-story house the diagonal floor for the second story does provide a stronger bracing effect than would a floor laid at right angles to the joists and hence is recommended because of this advantage, as well as for the final appearance of the finished floor.

Two 8d box nails are used in each board where it crosses each joist. Each nail should be placed about ¾ inch in from the edges of the board and driven until the head is tight. Joints are made on the center of a joist and should be scattered so that there are no more than two together on a joist.

To lay subfloor, proceed as follows (see Figure 24):

1. To lay subfloor straight, select good straight boards for the first continuous row. Start on the longest side of the house and carefully tack the board into place, flush with the outside face of the rim joist; then sight for straightness, correct if needed, and nail to place. This first board affords another opportunity to straighten any slight crookedness in the rim-joist work.

2. Continue with the second row, then the third row, etc. It is advisable to sight the front edge of the subflooring occasionally as the work proceeds, since framing lumber is not always milled perfectly true; some flooring boards may be narrower than others, and occasionally one may be found that will be narrower at one end than at the other. The main objective in laying subfloor is to keep the front edge as straight as pos-

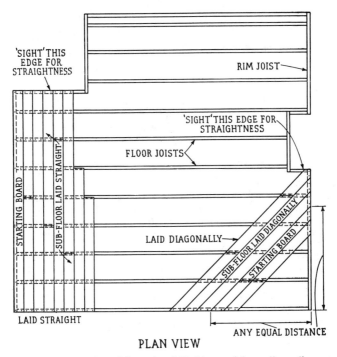

**Figure 24.** Starting subfloor when laid either straight or diagonally.

sible; this results in a craftsmanlike job. If the center becomes bowed, then spread the outside ends a trifle and vice versa.

3. The procedure for laying a subfloor at a 45-degree angle to the floor joists varies from the method described above only in locating, cutting, and nailing the first or "starter" board. To provide a starting board long enough to assure a straight edge and to make it at a 45-degree angle, measure from the starting corner of the building equal distances each way, say 8 or 10 or 12 feet. Stretch a line or make a chalk line from each similar measurement (hypotenuse of the angle) and cut and nail the first board at that place.

4. Work from both edges of this board, continuing as outlined above, occasionally sighting to keep the front or working edge as straight as possible.

All joints must be cut parallel to the joist; hence they will be cut to a 45-degree angle. It is common practice not to cut off the outside ends until the job is completed; then all subflooring protruding beyond the building is cut off flush with the outside face of the rim joists. A 45-degree miter square is excellent for marking the corner angle.

**Note:** If several men are laying the floor, it will speed up the job if the boards are nailed only occasionally, leaving the main nailing job until all the flooring is cut and held to place. Then a continuous nailing job can be done.
stopping to cut and fit the various subfloor boards.

A problem arises when linoleum is used in some rooms, such as the kitchen or bathroom, and hardwood is used throughout the balance of the house. Linoleum is approximately ⅛ inch thick; hardwood flooring varies in thickness from ⅜ inch to ½ inch.

When linoleum is used, the flooring under it must be smooth to provide a surface to which the linoleum can be cemented. Also, in a doorway between, say, the kitchen and the dining room, there would be a difference in height of ⅜ inch between the top face of the linoleum laid on the subfloor and a ½-inch hardwood floor laid on the subfloor. To make both surfaces flush, lay subfloor stock throughout the house as usual; then, *after* the house is plastered, lay ⅜-inch plyscore in rooms that are to receive linoleum. When ⅛-inch linoleum is laid on top of the plyscore, the surface will be exactly flush with the surface of the ½-inch hardwood floor.

The plyscore is fastened down with 3d 'ring-shank' or 'screw-type' nails which should be placed at 6″ o.c. along all edges and 8″ o.c. on the plywood face. Joints should be staggered and spaced 1⁄32″ to prevent buckles.

## GENERAL INFORMATION ON THE WALL UNIT

The wall unit includes all framing required to construct the walls and partitions of a residence, from the laying out and setting of the bottom plates to the doubling of the top plates in readiness to receive the ceiling joists. The jobs involved are the same irrespective of the number of stories in the house; that is to say, the second-story wall unit includes exactly the same jobs, methods of procedure, materials, and layout problems as does the first-story wall unit. Even in scaffolding the only difference is the greater height of the second-story scaffold, which in turn may require slightly larger scaffold members to obtain greater strength.

Specifically, a wall is constructed of upright members called studs and horizontal members called plates. Openings in a wall or partition require additional framing members called headers, top and bottom cripples, and trimmers. To obtain rigidity, braces are placed in a diagonal position at as nearly a 45-degree angle as possible. Horizontal members placed halfway between the floor and the ceiling are called *fire stops* (see Figure 25).

In trade language the term *wall* applies to the framing erected to form an exterior wall of a house. The term *partition* is used to designate the framing members which separate or "partition" one room from another. From the carpenter's viewpoint the work involved in either a wall or a partition is exactly the same.

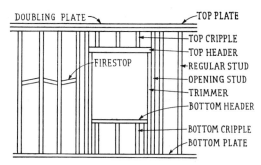

**Figure 25.** Framed-wall terminology.

For purposes of simplification the following discussion is divided into several topics: plate layout; stud and opening layout, including the story pole, corner construction, and inside-angle construction, erection of walls and partitions, including straightening, bracing (both temporary and permanent), and double plating; and exterior-wall storm sheathing.

## LAYOUT PROCEDURES FOR THE WALL UNIT

*Bottom-plate Layout.* The subfloor, it is assumed in the following description, has been completed, cleared of all blocks, and swept free of loose dirt.

The location of the bottom plates for all walls and partitions is determined by means of a tapeline and chalk line. All measurements on the floor plan must be carefully studied. Some architects show dimensions to the center of a partition; others may show dimensions *between* opposite walls (see Figure 26). In the latter case exact allowances must be made for the thickness of each wall. The plan may show a 6-inch wall thickness or a 4-inch partition thickness. Actually the thickness of a wall or partition is based on the width of the plate stock. A 6-inch exterior wall, to the carpenter, means $\frac{3}{4}$-inch siding or diagonal sheathing plus $3\frac{5}{8}$ inches, the exact width of a piece of S1E wall stud. A 4-inch partition, to the carpenter, means a $3\frac{5}{8}$-inch plate. Hence, when laying out for the location of the bottom plates, the carpenter cannot merely follow the plan dimensions, as a few "extra inches" will be picked up when a series of room dimensions are combined.

Outside-wall plates are located either flush with the rim joist or set in to allow for diagonal sheathing. The cross section of the wall shown on the blueprint will give exact details on this point.

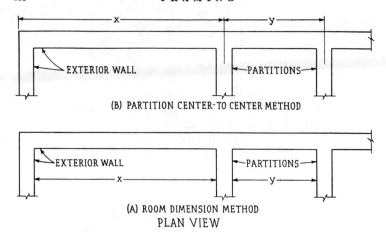

**Figure 26.** Alternate methods of dimensioning a floor plan.

Proceed as follows:

1. Using a carpenter's rule, measure in at each end of an outside wall a distance which represents the thickness of the wall (width of plate stock plus, perhaps, ¾ inch for sheathing thickness) and drive a nail at each point so located.

2. Fasten a line to each nail and strike a chalk line to represent the inside face of the bottom plate. This line provides one more opportunity to eliminate in the wall any crooked spots which may have developed when setting the floor joists.

Note: "Striking a chalk line" is not difficult if one proceeds in the proper manner. After the line is chalked it must *not* touch the floor until it is ready to be "snapped." Hence it is advisable to fasten the line to one of the nails; then chalk it, using a standard piece of blue or white chalk (made in the form of a semisphere). A second person should hold the line at its approximate center and lower it to the floor as the end man pulls the line taut and wraps it around the nail at the opposite end from which it was started. Do not let the line move after it is tightly drawn.

The center man then places his thumb on the line to hold it firmly to the floor; next he reaches out and carefully lifts the line off the floor and lets it snap to the floor. He then repeats the process for the other half of the line. The operation should result in a perfectly straight line which represents the inside face of the outside-wall plate (see Figure 27).

3. Repeat this operation for all walls.

4. Select the bottom-plate stock. This is the opportunity to use the poorer pieces of lumber, such as crooked ones or those with defects, for the material can be easily straightened as it is nailed, and defects do not

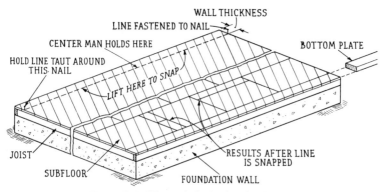

**Figure 27.** Striking a chalk line for plate layout.

affect the strength of the bottom plate, which does not actually carry a load but only transmits it from the studs to the floor.

5. Place the surfaced side of the lumber up, to simplify the stud markings, and spike the plate to the floor joists, using two 16d box nails at least every 32 inches o.c.

6. When these operations have been completed, the outside-wall plates are in position ready to locate the partition plates.

Note: The bottom plate is not cut out where it passes over the recess made in the floor joists to receive the exterior doorsill (see page 104) until after the stud layout has been completed. This procedure permits minor adjustments when laying out the door opening and also simplifies the stud layout, since the bottom plate will be continuous.

7. Select any partition desired, check the blueprints for dimensions, and, using a tapeline, locate one edge of the partition, measuring from the inside face of the exterior-wall plate.

Note: Plate-layout work is simplified if main partitions are located first, chalk lines made, and the plate stock "tacked" to place. Minor partitions are more quickly located by measuring from a main wall or main partition.

8. Measure with a carpenter's rule one-half the thickness of the partition, *using the dimension as shown on the blueprint,* and locate the *center* of the partition plate. For instance, if the partition is shown as 4 inches, then 2 inches will be the distance from the end of the tapeline measure to the center of the partition.

9. On each side of this center mark, locate a line equal in distance to one-half the width of the plate stock, in this case $1\frac{13}{16}$ inches (one-half of $3\frac{5}{8}$ inches). These two lines designate the exact position of the partition plate (see **Figure 28**).

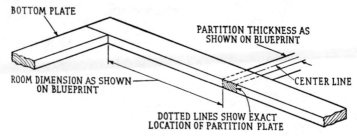

**Figure 28.** Marking for a partition plate.

10. From the *center line* continue this process and locate the remaining partitions on this outside wall. Then repeat this procedure on all the other exterior walls.

Note: The proof of the accuracy of the layout will be shown in the last measurement made when the opposite side of the building is reached. If any discrepancy is found, tapeline measurements have not been accurate, the blueprint measurements are not correct (this will occasionally happen even on plans which have been most carefully drawn), or the last measurement may be "long" because of a difference between the partition dimensions shown on the blueprint and the actual dimensions of the lumber. It is sometimes necessary to check the separate room measurements, plus the wall and partition thicknesses, and see if they add up to the over-all plan measurements.

Slight discrepancies of 1 or 2 inches in a plate layout are left unchanged if they are not important or, if certain room measurements, such as the space for a closet, appear to be crowded; then any gain is added at that place. It is therefore advisable to make a "dry-run" layout from wall to wall before deciding on the final location of each partition plate.

11. Stairway partitions must be very carefully located since any *major* change in location will affect the stair layout.

12. Snap chalk lines to represent one face of all partitions.

13. "Tack" the plate stock to the subfloor, being sure to place it on the correct side of the chalk line, as shown by the two lines made on the inside edge of the exterior-wall plate (see Figure 28).

Note: At this stage of the work, partition plates should merely be "tacked" to the subfloor with 8d box nails, driven in only part way, in case it is found necessary to move a plate. Final nailing is not done until after doorways are located, as the bottom plate must be cut and removed at every door opening.

All partition plates are then spiked or toenailed to the floor joists. One 16d box nail is used for each joist, but the nails are staggered so that both edges of the plate are nailed. Nails are *not* driven in the center of the plate. If plates are parallel to the joists then toenail, using 8d box or 8d common every 2 feet.

***Stud and Opening Layout, Basic Principles.*** Before the details of how to lay out for studs and window and door openings are given, several basic principles need to be explained. First, it may not be apparent to the beginner, at this point of the discussion, why it is worthwhile to start all stud layout from one side of the building or from either the front or the rear, but never from both ends or both sides. The reason is that this procedure automatically locates the ceiling joists as they are placed directly above the wall studs. On a correctly laid-out job all regular studs should "line up" when you sight through the building after the walls and partitions are framed and erected.

A second basic principle to follow is uniformity of spacing. Studs are spaced 16 inches o.c. to conform to the lath measurements. Studs so spaced are known as regular studs. Laths are made in lengths which are multiples of 16, such as 32, 48, or 96 inches.

The third principle of stud layout pertains to the location of the door- and window-opening studs. The 16-inch o.c. spacing is continued from one wall or partition end to the other; then the opening studs are located. Two opening studs are usually required, one for each side of each opening. Sometimes a regular stud (see Figure 25) will also serve as an opening stud. Occasionally an opening may be framed with two regular studs.

Fourth, it is very important to start the stud layout correctly to avoid lath waste. Four studs will form three 16-inch o.c. spaces, as a 4-foot piece of plaster lath spans four studs, centering on each of the end studs (see Figure 29). However, when an inside angle is involved, the lath must start at the angle and center on only one end stud. To be definitely sure that the stud spacing is started correctly: (1) measure from the inside angle a distance of 4 feet to find the center of the third stud (see Figure 30); (2) mark 1 inch on each side of the center mark to represent the two sides of this stud; (3) using the steel square, as was done for the joist layout (see Figure 9), lay out the studs in both directions from this stud.

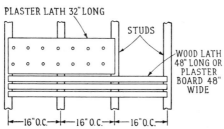

**Figure 29.** Stud spacing in relation to length of lath.

**Note:** If the house is to have a plastered exterior, it is considered good trade practice to start the layout for the exterior-wall studs from the *outside* corner of the building and not the inside corner of a room, as outlined above (see Figure 30). The procedure of measuring 4 feet from the corner to locate the center of the third stud is still applicable. Also, opposite walls should be laid out from the same side or end of the building so that the ceiling joists, which are located directly above each stud, will be parallel to the exterior walls.

Either method of stud layout, that is, measuring from the outside corner of the building or the inside corner of a room, will eliminate the waste of lath materials.

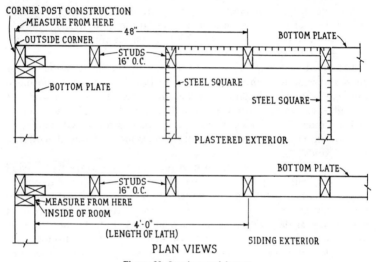

CORNER POST CONSTRUCTION

**PLAN VIEWS**

**Figure 30.** Starting stud layout.

*Outside-corner Posts.* Before describing the stud-layout procedure in detail, it is necessary to illustrate and describe the stud construction for an outside corner and for an inside angle. Corner posts must be constructed to form all of the outside corners of a building. They are made by assembling three studs as shown in Figure 31.

The method of construction shown is considered one of the best because the thickness of the stud does not affect the width measurements of the post. The straightest studs should be selected for corner posts as the finished assembly must be straight. It is customary for the carpenter to cut the required number of studs to frame all the walls and partitions (the number can be counted on the plate layout) and then to select the best and straightest ones for all corners, angles, and opening studs.

Corner posts can be "made up" in advance of raising the walls, for they can be made to fit any corner by turning them end for end.

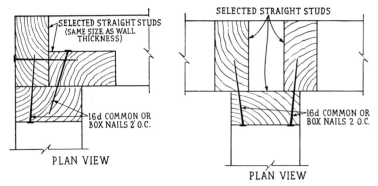

**Figure 31.** Stud construction to form an outside-corner post.

**Figure 32.** Inside angle on partition posts.

***Inside-angle Posts.*** Inside angles in a frame wall are formed by the 90-degree intersection of a partition with an outside wall or with another partition. The framing at this intersection must be constructed to make the post definitely tie the wall and partition together and also provide lath nailing in both directions (see Figure 32). Occasionally an opening header will necessitate a different post assembly; a good layout man can quickly tell what the framing conditions at every intersection will be by looking at the stud layout on the bottom plate.

In the same manner as described above, posts for inside angles are constructed by selecting the best and straightest stock.

**Note:** It is possible to take care of the crooked studs by placing them in a straight wall in such a way that the crookedness can be offset or relieved. See page 132 for detailed instructions on how this is done.

Outside posts and inside-angle posts are assembled by spiking them together with 16d box nails, as shown in Figures 31 and 32. One nail should be placed for at least every 18 inches of height. Slanting the nails in different directions increases their holding power.

***Stud Layout on Lower Plate.*** Keeping in mind the basic layout principles and the description of corner posts and angle posts, proceed as follows to lay out the stud locations on the bottom plate:

1. Lay out all corner-post and angle-post stud locations on the bottom plate, marking carefully to conform to the exact shape of the assembled posts.

2. Select the exterior walls from which the layout is to be started.

3. Locate the center of the third stud by measuring 4 feet from the inside angle and draw the two lines which represent the stud sides.

4. Using a 16-inch-tongue steel square (be sure it *is* 16 inches), place as shown in Figure 9 and make two lines for each stud, continuing until the end of the wall is reached. Whenever a partition is encountered, continue to space the studs on the 16-inch o.c. spacing. Do *not* break this spacing continuity. Some adjustments will need to be made when a regular stud comes at exactly the same location as one of the studs forming the partition post.

5. Repeat this operation on the *opposite* wall, being sure to begin the layout at the same end of the building from which the first wall layout was started.

6. Continue this operation on the exterior walls running in the opposite direction.

7. Lay out all partitions, starting from a partition post and again measuring out 4 feet to locate the center of the third stud.

***Window and Door Layout, Horizontal Measurements.*** The next layout job concerns the door and window openings. It is customary for architects to show window locations on the floor plan. Dimensions are given to the center of each window unless a window is to be framed directly into an outside corner (see Figure 33).

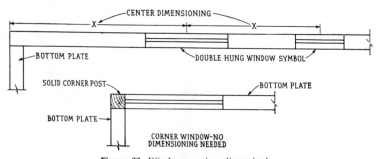

**Figure 33.** Window-opening dimensioning.

If a dimension indicates that a window is to be centered on the wall of a room, find the exact center of the wall *as laid out,* even though this dimension may vary slightly from the floor-plan measurement. As stated earlier, a few inches of length are usually "picked up" on any plate layout because of the difference between the architect's wall- or partition-thickness measurement and the job dimensions which are developed from the actual width of the lumber used for plates; hence the latter dimensions must be used for the window layout.

Unless a wall space has been planned to fit a piece of furniture exactly, door-opening locations are rarely dimensioned on the floor plan, but their location is determined by scaling the blueprint. A door opening that is adjacent to a stairway must be carefully located in its exact relation to the stair measurements, which must be followed very closely.

The width of a window or door opening between opening studs is definitely conditioned by the trimmer studs and the thickness of the

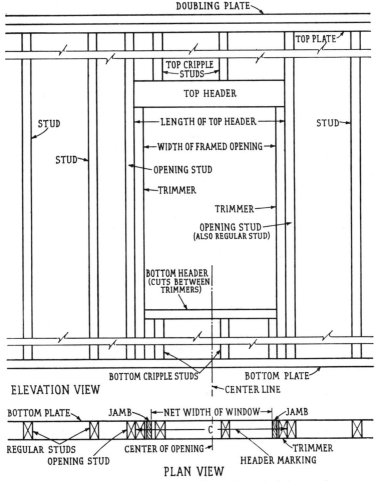

**Figure 34.** Factors which determine width of a framed window opening.

window or doorframe jambs. The *kind* of window that is double-hung (slides up and down) or casement (swings in or out) may also make some difference in the width of the framing space between two opposite opening studs (see Figure 34).

**Note:** The section on window and doorframe construction in Chapter 7, pages 225–234, should be carefully studied at this time.

Each opening in a framed wall requires a horizontal member across the top to carry the load. The size of the top header is based on the width of the opening; building ordinances give required dimensions for various spans.

Top headers are supported by trimmers which are cut to fit snugly between the underside of the header and the bottom plate (see Figure 34).

See Page 300 for sliding door framing.

Bottom headers are made from lumber of the same size as used for the studs as they do not carry any load.

To lay out a window or door opening on the bottom plate, proceed as follows:

1. Read the dimensions on the floor plan and locate the same point on the plate. If the measurement is one which designates the center of the opening, it is good trade practice to mark the center point with a large C passing through the center line (see Figure 34).

2. Read the width dimension of the opening and divide by 2. For instance, if the window is marked 2′6″, this measurement is divided by 2 to obtain 15 inches.

3. Measuring in opposite directions from the center, mark a distance equal to one-half the width of the window or door. These marks represent the inside face of the finished window or doorframe.

4. Determine how much space will be needed to allow for the thickness of the frame jamb stock. This will vary from ¾ inch to 1⅝ inches, depending on the frame detail.

5. Measure from the mark which represents the inside face of the jamb stock the amount determined in step 4 and mark. This mark actually represents the outside face of the frame jamb.

6. Determine how much room will be needed to give enough clearance for the frame to be set easily and correctly. No set amount can be given; some clearance must be allowed to take care of any framing discrepancies —variations in the thickness of the framing material or a slight "out-of-plumbness" of the trimmer studs.

**Note:** While the spring sash balance for the double-hung window is now in common use for residential buildings, sash weights are still required for such jobs as school buildings.

Sash weights require weight pockets to slide in, which must be planned for when the width dimensions of a window opening are laid out. The size of the pocket will vary, depending on the size of the window. The larger the window, the heavier the weight required. For a window of ordinary size, such as 3'0" by 4'6", a 2-inch weight pocket must be allowed.

7. The marks finally made in step 6 represent the inside face of the trimmer, and 2 inches more is now needed to allow for its thickness. The mark made for thickness also represents the inside face of the opening stud. This last measurement is the all-important one, for the distance from the face of one opening stud to the opposite stud represents the exact length of the top header. Each opening should be marked with arrow points, as shown in Figure 34, to provide a reference method for locating each opening quickly. Headers are cut from arrow point to arrow point.

8. Adjustments in the regular stud layout which conflict with the opening stud must now be made. Sometimes the opening stud and the regular stud can be one and the same piece.

Occasionally an opening can be moved horizontally, provided that it is not centered on the wall. In every case it is important not to lose or move the location of the regular stud, for the continuity of the 16-inch stud spacing must be retained to assure correct spacing for the lath.

***The Story Pole.*** The plate layout covering horizontal measurements of a door or window opening is now completed. The next step concerns the vertical measurements. An entirely new set of measurement factors must be planned for, including the over-all height of the wall, the kind of plasterboard used, the location of the top line of all the openings, thickness of the frame head jamb, clearance allowance

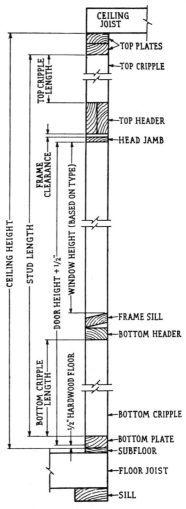

Figure 35. Completed story-pole layout.

above the head jamb, and the thickness and slope or pitch of the frame sill. These measurement factors may seem to make the vertical layout a complicated job, but it is simplified greatly by means of a story pole (see Figure 35).

A story pole is actually a vertical full-sized cross section of the wall, from mudsill to ceiling joist, showing the distance between the top and bottom plates and between the top and bottom headers of all openings. These framing members are laid out and marked in their exact vertical position, so that it is possible to determine the exact length of all full-length studs and all top and bottom cripples (see Figure 34).

A piece of 1 by 4 S1S stock is selected for the pole, the length of which is a few inches more than the total height of the exterior wall, including the sill and floor joists. A vertical cross section of the wall is usually shown on the blueprint; sometimes the ceiling height is marked on the elevation sheet.

The carpenter has a little freedom in determining the exact height of the wall, for he adjusts the height on the basis either of the material used on the exterior walls or of the material used on the interior, which may be lath and plaster or plasterboard (known as dry wall).

The height of all openings is generally determined by the door height, which is usually 6'8''. On some jobs, however, doors are 6'6'' high, and in more pretentious homes doors may be 7'0''. To give a pleasing appearance to all the house elevations, the tops of the windows are usually in line with the tops of the exterior doors—hence the importance of knowing what door height is shown on the plan. There may, however, be some windows that are not located vertically according to this rule; in this case the blueprints will dimension the location of the top line of the window.

If interior dry-wall construction is specified, the ceiling measurement, from subfloor to ceiling joist, must be no more than 8'1''. Plasterboard is manufactured in 8'0'' lengths; the 1-inch clearance allows for the ½-inch thickness of the boards (when nailed on the ceiling) and provides ½-inch clearance at the bottom, so that no difficulties are encountered as the large sheets of plasterboard (4'0'' by 8'0'') are handled and nailed to the wall.

**Note:** If the floor-to-ceiling measurement were 8'4'' (as is often specified), there would be 4 inches of uncovered wall space that would have to be filled with waste pieces of wallboard accumulated from around the openings. This would not be economical. Also, it would result in a very narrow wallboard nailing surface at the top and bottom plates.

For siding exteriors it is a great advantage to lay out on the pole or "rod" the position of each siding board, starting with the bottom board and marking the net-coverage width of each board until the top plate is reached. A 6-inch piece of siding actually covers 5 inches of vertical space; an 8-inch board will cover 7 inches; this difference is due to the ½-inch rabbeted edge on the bottom and the loss in milling, which amounts to ½ inch (see Figure 36). By adjusting the wall height a little, it is sometimes possible to eliminate one row of siding boards for the full perimeter of the house.

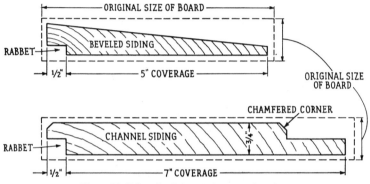

**Figure 36.** Siding detail in relation to wall height.

The length of the stud material is another factor which must be considered when laying out the story pole. The thickness of one bottom plate and two top plates totals $4\frac{7}{8}$ inches, assuming that each plate is $1\frac{5}{8}$ inches thick. Stud material is cut in even-foot lengths, such as 8′0″ or 9′0″. To avoid as much end waste as possible, it is desirable to make the height from floor to ceiling on the basis of maximum stud length plus plate thickness (assuming that the other conditions described above can also be met). If an 8′0″ length of studding is to be used, it should be figured as 7′11″ to allow for the squaring of each end of the piece, as it cannot be assumed that each piece of studding lumber will be perfectly square on each end.

Finally, building ordinances must be considered since they indicate minimum ceiling heights.*

RIBBON CONSTRUCTION FOR BALLOON FRAMING. The preceding description refers to one-story walls and partitions that are framed from the bottom plate to the top plate, with the ceiling joists resting on the top plate. On a two-story house, framing is sometimes done with the studs on the outside walls continuous from the first floor to the second-floor

* The house requirements of the County of Los Angeles, for example, include the following regulation: "No room shall be less than 7′–6″ in height, in at least 50 percent of the floor area of the entire house."

ceiling joists. This is known as balloon framing; the one-story-at-a-time method is called platform, or Western, framing (see Figure 37).

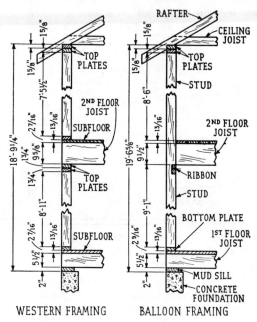

**Figure 37.** Balloon and Western (platform) framing.

One-story flat-roof houses are sometimes built using the same type of ribbon construction method. In this case, however, the studding need

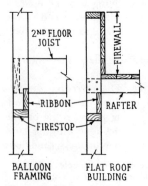

**Figure 38.** Ribbon construction.

only be increased in length so that it extends from the floor to the roof plate instead of from floor to ceiling (including plates), as in platform framing. In an average house the wall above the ceiling-joist line need not be much more than 3'0" or 3'6". In trade language, the exterior wall above the roof line is called a "fire wall." The ribbon method is also used in building flat-roofed residential garage buildings.

This framing method is illustrated in Figure 38. Note the extra piece of fire stop (see page 140). A piece of 1 by 4, called a ribbon, is dapped or notched into the inside edges of all wall studs that are to support the ceiling structure. The top edge of the ribbon is made the same height as the partition so that the ceiling-joist-bearing surfaces will be level.

On two-story houses the second-story joists are sometimes notched to fit over the ribbon to provide additional holding power between the walls of the house.

PLATFORM FRAMING. Platform framing is increasing in popularity throughout the United States because it permits framing one story at a time, each succeeding story being constructed on top of the preceding framing work. Obviously, if a building was to be, say, three stories in height, it would not be possible to secure studding long enough to make a continuous stud from the sill to the third-story top plate; even if this were possible it would not be practical to try to construct the building in this manner. Hence it is becoming the general practice to follow platform-frame construction methods.

The story-pole layout is done in the same manner for either platform or balloon framing, except that the layout from the top-plate line must be carried on to the top of the fire wall if the latter type is used. The blueprints will indicate the height of the fire wall.

STORY-POLE LAYOUT. Dimensions given for the story-pole layout shown in Figure 35 are for purposes of illustration only. As emphasized above, final vertical measurements are determined on the basis of ordinance requirements, blueprint measurements, kind of finish to be used on either the exterior or interior, and the length of studding material ordered. Obviously not all these requirements can be met at any one time; adjustments must be made on the basis of the job conditions. Keeping these considerations in mind, as well as the foregoing general discussion of the story pole, proceed to lay out the story pole as follows:

1. Select a suitable 10-foot piece of 1 by 4 S1S.
2. Smooth the edges if they are rough.
3. To keep the piece from sagging, put it on top of a 2-inch plank laid on a pair of horses.
4. Begin at the left end and make two lines to represent the mudsill.
5. Next draw a line to represent the top edge of the floor joists.
6. Draw in another line to represent the face of the subfloor.
7. Now draw in the bottom-plate thickness.
8. Measure from the subfloor a distance equal to the floor-to-ceiling height, as predetermined from the blueprint, the wallboard measurements (8′1″, as explained above), or the number of siding boards required to cover the exterior wall from sill to top plate. This vertical measurement is the all-important one since it determines, once and for all, the framed heights of all walls and partitions. Make a line at this

point. Then measure "down" and make two marks to represent the two top plates.

9. Read the specifications to determine the thickness of the finish floor and make a line above the face of the subfloor to represent the face of the finish floor.

10. From the finish-floor line measure a distance equal to the door height shown on the blueprint (usually 6′6″, 6′8″, or 7′0″). Then add ½ inch for rug clearance, and mark. This line represents the underside of the finish doorjamb.

Note: Rug-clearance allowance permits the hanging of a door without trimming off the bottom edges. If the vertical door measurement is made "net," it is necessary to saw off ½ inch from the bottom of the door to make it swing clear of the rug. For exterior doors this ½-inch space is filled by a piece of threshold and, in cold climates, by suitable weatherproofing material.

11. Next make a line to represent the top side of the doorjamb. This measurement ranges from ¾ inch for a 1-inch jamb to 1⅝ inches for a 2-inch jamb.

12. Allow at least ½ inch for head-jamb clearance, and mark. This line represents the bottom edge of the top header. Determine the width of the header stock, measure, and mark. This line represents the top edge of the top header.

Note: Narrow openings may require only a 4 by 4 header, made of two pieces of 2 by 4 spiked together and *set on edge*. Wide openings may require a header as large as 4 by 12, made from two pieces of 2 by 12. Plans and specifications or building ordinances are quite specific on this point.

13. Now start from the mark which represents the *face of the finished head jamb* and measure *toward the bottom* of the pole to locate the position of the bottom header. A full knowledge of window-frame construction is needed to do an accurate job as there are several factors which cannot be overlooked: (*a*) whether the sash opens in or out or slides up or down, (*b*) the pitch of the window sill, and (*c*) the sill clearance required to make the frame setting easy.

A study of the section on window and door frames in Chapter 7 is quite essential. Suffice it to say here that the exact position of the bottom header must be determined and that there may be several heights to mark, since windows in an average house vary in length.

It is essential that a cross section of the sill laid out to the correct slope or pitch be made on the story pole. A T bevel is very useful in "picking up" the slope from the blueprints, or the T bevel can be set to the correct slope by placing it on the sill stock itself, if a piece is available.

SUMMARY. The pole or rod, when laid out as described above, furnishes the carpenter with all the measurements he will need in cutting full-length studs, top cripples, and bottom cripples. If top headers of different widths are required, the top-cripple studs will be of different lengths.

It is now possible to count the exact number of pieces required for the various kinds of studs—opening, top cripples, and bottom cripples— and to know the exact length that they should be cut. Headers can be cut to exact length as measured on the bottom-plate layout and spiked together ready to assemble in the wall. Hence the next topics will pertain to top-plate layout and how to make and use a stud-cutting box.

*Top-plate Layout.* In all residential construction two plates are used at the top of a framed wall or partition. The first, or lower, plate must be laid off exactly to duplicate the stud markings on the lower plate. The second, or top, piece is called a *doubling plate.* Its purposes are to give the wall rigidity and additional strength for carrying any ceiling or roof load, to provide a satisfactory method of tying together a butt joint in the lower plate, and, lastly, to make possible a good half-lap joint at every intersection of walls and partitions (see Figure 39). The doubling plate requires no stud layout.

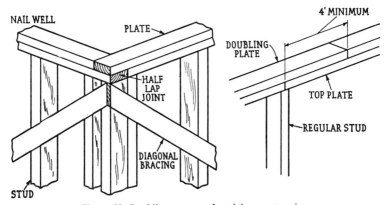

Figure 39. Doubling- or top-plate joint construction.

The other (lower) plate, usually referred to only as the "top plate," may need to be joined on a long wall, since 20 feet is the maximum length usually carried in stock by a lumberyard. (Longer lengths necessitate a special order and hence cost more.) A wall or partition longer than 20 feet will require a butt joint for the plate, which must be made in the center of a stud. At no time should this butt joint be made over an opening, lest difficulty be encountered when raising the walls. This procedure is explained on page 136. The doubling plate serves to tie the butt joint together, and it must be joined at least 4 feet away from the top-plate joint.

Lumber selected for top plates must be straight and free from serious defects. It is a basic principle of house framing that it is much easier to select straight stock for major framing items, such as plates, than to attempt to straighten the item. Plate stock may have a slight bow when sighted along its edge; this can be compensated for by placing the bow of the doubling plate in the opposite direction from that of the lower piece. Spiking two pieces together which have opposing "bows" will automatically straighten them.

Since two 16d box nails must be driven through the plate into each stud, the plate is less likely to split during this process if it is constructed of straight-grained wood. Doubling plates are nailed alternately from each edge, and the 16d box nails are spaced about 4 feet apart.

To lay out the top plate, proceed as follows:

1. Select straight and straight-grained plate stock, using lengths which will eliminate all possible butt joints.

2. Determine the studs on which a butt joint must be made.

3. Cut the top-plate stock to lengths which correspond with the bottom plates, or which will join over a stud, continuing until all wall and partition top plates are cut.

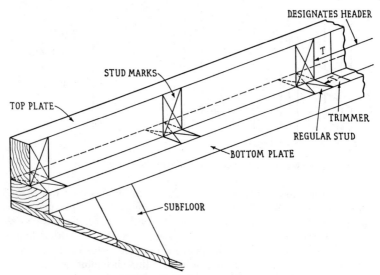

**Figure 40.** Marking top plate.

4. Select the face of the plate which will be placed on top of the studs. Then turn the plate on edge in such a position that it is possible, by means of a try square, to transfer all stud markings to it, duplicating all stud marks already made on the bottom plate (see Figure 40).

The doubling plate, requiring no marks, is cut and nailed into place after the frame is raised. It should, however, be carefully selected, stacked in a way that will help hold its shape, and labeled to prevent usage on some other part of the building.

The plate layout for walls and partitions is now complete. The next job is to cut studs and headers.

**Note:** There is no one procedure to follow when framing the walls and partitions of a building. The various steps involved are discussed here in a logical order: (1) layout to determine the number, length, and location of the many pieces required for a frame residence; (2) cutting the pieces to length in readiness for assembling; (3) assembling; and (4) raising the wall and partitions. Practical methods of accomplishing each step are described.

## CUTTING FRAMING MEMBERS FOR THE WALL UNIT

*Cutting Studs and Top Cripples.* Stud cutting must be accurately done; cuts should be perfectly square and the lengths exact. Modern contractors use power hand tools to achieve this result. It is necessary to construct a flat table on which the stock can be laid when using the power crosscut saw, unless the saw is equipped with a metal frame designed for this purpose.

The man who is constructing his own home and who does not own a power saw must use a different cutting method. This would also be true for the contractor on jobs in rural districts where it is too costly to run a temporary power line.

CONSTRUCTING A STUD-CUTTING-BOX. If the use of a power saw is not feasible, it is necessary to construct a *stud-cutting box,* which is a formidable name for a very simple device (see Figure 41). To construct the box, proceed as follows:

1. Toenail together two 2 by 4s, each 16 feet long, to make a 4 by 4. This gives the box rigidity.

2. Lay the 4 by 4 on a pair of sawhorses, which should have been placed adjacent to the lumber pile and leveled to prevent racking.

3. Near the center and on opposite sides of the 4 by 4, nail on, vertically, two pieces of 1 by 6 so that they project at least 8 inches above the face of the 4 by 4. They should extend a few inches below the box to prevent splitting. The space between the upright pieces should be in-

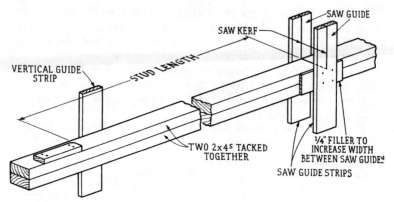

**Figure 41.** A stud-cutting box.

creased a little to make it easier to place a 2 by 4 between them. A ¼-inch-thick strip of wood can be nailed on the side of the 4 by 4 to increase the space.

4. Using a square, make a horizontal line on the 4 by 4 and a vertical line on the 1 by 6; then carefully cut a saw slot on these lines. Be sure the line is accurately cut for squareness, for this saw cut will determine the accuracy of the end cut of each stud.

5. Measure to the left from the saw cut a distance equal to the length of the studs, as shown on the story pole (the distance between the bottom and top plates). Square a line at this point and nail on a piece of 1 by 4 to act as a stop. The inside corner of the stop is cut at an angle to form a recess to receive any accumulation of sawdust. This is important, for otherwise the studs would be cut in varying lengths. The "butting edge" of the stop must at all times be kept free from dirt.

6. Nail on a guide strip to form a back to the box. This strip permits lining up the stud material parallel to the box. The box is now ready for use.

7. To adjust the box to permit cutting the top-cripple studs, simply nail on at the correct distance another stop block.

**Note:** Never move the full-length-stud stop block since additional studs may be needed as the job progresses.

USING A CUTTING BOX. To cut studs by means of the cutting box, proceed as follows:

1. Pick up a piece of stud material and place it in the box. Test the end which is to be butted against the stop block for squareness; if necessary, cut this end using the saw cut in the 1 by 6 as a guide. This eliminates the necessity for making a mark to saw by.

2. *Gently* place the squared end of the stud against the stop block and cut the piece to its correct length. Be sure *not* to bump the block; this would eventually cause it to move and result in studs of varying lengths.

3. As the studs are cut, sight each one and place the straight ones in one pile to make it easy to select pieces for corner posts and opening studs.

4. Count on the bottom-plate layout the number of top cripples required between arrow points at each opening. Measure their length on the story pole (there may be several different lengths because of differences in the width of top headers) and nail on a stop block at the correct location. Crooked stud material should be used for the cripples, for cutting a crooked piece of lumber into several short ones automatically eliminates the bow.

Note: Bottom cripples are not cut until after the wall is raised since the bottom header is likewise cut and fitted at that time. Bottom cripples can then be cut to fit snugly between the bottom header and the bottom plate. This procedure provides one more method of securing an excellent framing job. Bottom headers may vary slightly in thickness; if the bottom cripples have been cut in the cutting box, a few may be found to be too short or too long.

*Cutting Top Headers.* The length of top headers is indicated on the bottom-plate layout. If the layout work has been done correctly, an arrow point should designate each end of each header; hence measuring from arrow point to opposite arrow point gives the length to cut that particular header.

The top header may be constructed out of two 2 by 4s spiked together (see Figure 42). These are placed on edge; therefore, the combined thickness of two pieces of header stock must not be more than the exact thickness of the wall or partition. This is very important. Since framing lumber is usually S4S stock, it will be found that the net thickness of a 2 by 4 will be $1\frac{3}{4}$ inches; twice this amount equals $3\frac{1}{2}$ inches—slightly under the net width of a 2 by 4, which is $3\frac{5}{8}$ inches. The headers are usually nailed flush with the exterior face of the wall.

The depth of the header (vertical measurement when in the wall) is based on the width of the opening. The blueprints, story pole, and lumber list should be checked to ensure that the correct size or header stock is selected for each opening. Top headers on wide openings vary from a 4 by 6 to a 4 by 12 (either a solid piece or built-up).

After the top headers are cut, they are spiked together with 16d box nails, which should be placed $\frac{3}{4}$ inch in from each edge and approximately 4 inches from each end of the header.

Bottom headers are cut *between* the trimmer studs, which extend from

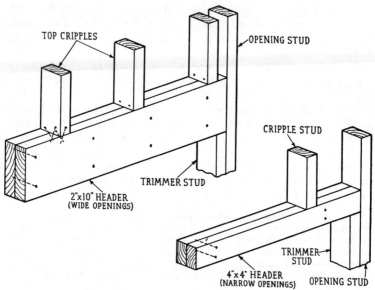

**Figure 42.** Top-header construction.

the underside of the top header to the bottom plate. To secure a good, tight framing job, trimmers and bottom headers are cut and nailed in place *after* the walls are raised. This method compensates for any discrepancies which may have occurred, such as opening studs that are nailed exactly to the layout line or slight differencess in the thickness of the trimmer studs.

## ASSEMBLING WALLS AND PARTITIONS

A number of preliminary jobs have now been completed in readiness to assemble the framed parts: bottom and top plates have been cut and laid out; corner and partition posts have been assembled; full-length studs and top-cripple studs have been cut; and top headers have been cut and nailed together.

The next step is to determine the order of framing, which is based on the order of raising. If the inside walls were raised first, then there would be no area on which the outside walls could be framed in readiness for raising. Exterior walls are therefore framed first and then raised to provide a working area on which other walls and partitions can be assembled.

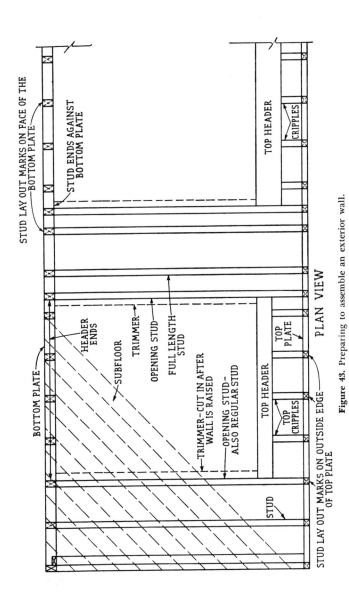

STUD LAY OUT MARKS ON FACE OF THE BOTTOM PLATE

STUD ENDS AGAINST BOTTOM PLATE

TOP HEADER

CRIPPLES

BOTTOM PLATE

HEADER ENDS

SUBFLOOR

TRIMMER

OPENING STUD

FULL LENGTH STUD

TRIMMER–CUT IN AFTER WALL IS RAISED

OPENING STUD– ALSO REGULAR STUD

TOP HEADER

TOP CRIPPLES

TOP PLATE

STUD

PLAN VIEW

STUD LAY OUT MARKS ON OUTSIDE EDGE OF TOP PLATE

**Figure 43.** Preparing to assemble an exterior wall.

The subfloor should be cleared of all debris and top headers placed at the opening for which they have been cut. Now proceed as follows:

1. Mark the location of the top cripples by placing the header in position on the bottom plate and transferring the stud markings from the plate to the header.

2. Toenail the top cripples to the top header, as shown in Figure 42. Use four 8d box nails per stud.

3. Select the opening studs for each opening. These should be carefully selected for straightness and soundness.

4. Measure down from the top end of each opening stud a distance equal to the length of the top cripples and square off a line. This will locate the top edge of the header.

5. Determine which wall to frame first and lay all studs and headers in their proper position with the bottom end of all full-length studs butting against the bottom plates (see Figure 43). Each stud should be sighted when placed in position and the crown edges of the studs reversed, so that for each alternate stud the crown edge is "up" and for the intervening studs the crown edge is "down." This procedure will help make a true wall surface, for each stud acts against the one next to it, and the final effect is to give a true, straight surface.

**Note:** Do not use studs with excessive crowns or bows unless it is absolutely necessary to do so; but if they must be used, mark them to designate that they

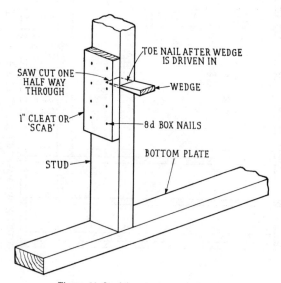

**Figure 44.** Straightening a crooked stud.

are crooked. *After* the wall is raised, straighten them by making a saw cut near the center and about two-thirds the way through the stud. Then drive a wedge in the saw cut, continuing until the stud is straight when tested with a straightedge. Then, using several 8d box nails, nail on a cleat or "scab" to strengthen the stud where the saw cut was made (see Figure 44).

6. Place the top plate in position and spike each stud to place, using two 16d box nails to each stud. Be sure that the opening studs are placed to line up with the marks against which the arrow points have been made.

7. Raise each wall as soon as it is framed (see below). Continue with the assembling process until all walls and partitions are framed.

## R A I S I N G  W A L L S  A N D  P A R T I T I O N S

The number of men on the job has considerable bearing on the problem of raising walls and partitions. The exterior walls of a large residence must, of necessity, be raised in sections. The simplest rule is this: "Frame as much of a wall section as can conveniently be raised with the manpower on the job."

The bottom ends of the studs should be placed against the bottom plate as shown in Figure 43 to prevent them from slipping as the top plate is lifted into the vertical position. The accuracy of the stud cutting will now be evident; studs cut exactly square will stand erect in a plumb position with little danger of tipping while temporary braces are nailed on.

Temporary braces are made from 1 by 4 or 1 by 6 material; roof sheathing is often used since it is available for such temporary uses. Braces should be long enough to provide as nearly a 45-degree angle as possible. To illustrate, an 8-foot wall will require a brace length equal to the diagonal of an 8-foot square, approximately 12 feet (see Figure 45).

While two carpenters hold the wall in place, a third carpenter should toenail the bottom ends of the main studs, that is, the corner studs and opening studs. Then a brace is nailed from the top of one of the studs across the outside face of the studs to the face of the header joists. The wall should be plumbed accurately before nailing the bottom end of the brace. Fasten another brace at right angles to the wall and to the bottom plate of a nearby partition. It is often necessary to put on temporary braces and remove some of them each time a wall or partition is raised. Each succeeding wall or partition helps to support the structure as the raising process continues.

Four 8d box nails are toenailed into the bottom end of each stud. It is important that these nails be driven "home"; that is, the head of

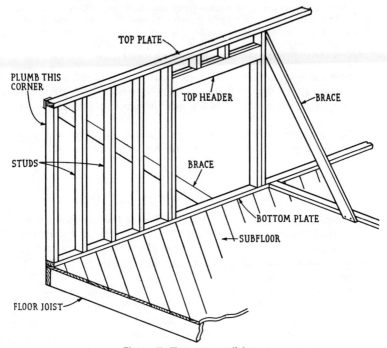

**Figure 45.** Temporary wall braces.

the nail should be driven until it is actually flush with the wood. In this way the holding power of the nail is increased considerably.

Plumbing the four corners of a building that has been properly laid out will automatically cause all partitions to be plumb also, provided that the location of every partition was marked on the edge of the top plate prior to raising the walls.

After all walls and partitions have been raised, additional braces are placed at strategic locations to prevent any movement of the frame. Brace nails should be left with the heads sticking out far enough to permit their withdrawal by means of a claw hammer.

A careful check should now be made to see that all studs are properly toenailed to the bottom plate.

In summary, to raise walls and partitions, proceed as follows:

1. Select one of the outside walls and raise it to a vertical position.
2. Toenail several of the key studs to the bottom plate.
3. Plumb the ends of the wall section and nail on temporary braces.

4. Select another wall and continue this process until all exterior walls are raised.

5. Assemble frame and raise the inside partitions, selecting the main ones first. Short partitions are usually left until the last, but care must be taken to ensure that it is possible to raise these short partitions.

**Note:** There is no set order for raising a frame, except that the outside walls must be raised first. Inside partitions are raised in any convenient order; the longer ones require more room for assembly and are therefore usually raised first.

6. Double-check the plumbness of each corner by testing opposite ends of the same wall. Both ends should register exactly plumb if the layout and cutting work have been accurately done.

7. Nail on additional braces to make the raised frame very rigid and to prevent it from getting out of plumb as the ceiling unit is framed. An ounce of prevention is worth a pound of cure.

## FINAL STEPS IN WALL FRAMING

*Straightening a Wall.* After a building frame is raised and plumbed, the next step is to straighten the walls and partitions. This is done as follows:

1. Nail on a ¾-inch block at the outside top end of the two corner posts which form the ends of the wall to be straightened.

2. Stretch a line tightly from block to block. This provides a line that is parallel to the wall and yet does not touch the wall (see Figure 46).

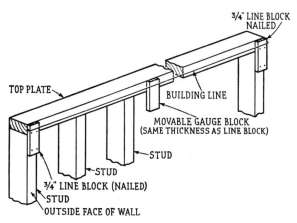

**Figure 46.** Straightening a wall by means of a chalk line.

3. Use a third block, the same thickness as the two blocks nailed onto the face of the studs, as a gauge between the line and the top plate, and move the top plate accordingly until the block just slips into the space. To hold the plate to this exact position, nail on a temporary brace. This may require that a piece of plate stock, say 2 feet long, be tightly nailed on top of the subfloor to receive the lower end of the brace.

Continue the "line-and-block" method on all walls and partitions until each one is perfectly straight. This method may not be necessary for short walls, however, since they can usually be straightened by sighting.

**Doubling-plate Nailing.** The doubling plate is now spiked on top of the top plate. Butt joints should be at least 4 feet away from the top-plate butt joints. The joint where a partition joins a wall is made in the reverse order to make a half-lap joint effect (see Figure 39, Page 125).

The doubling plate should be spiked so that there is always a "free end" which can be moved slightly from side to side to make the edges of the two plates perfectly flush. The nailing is done by straddling the top plate, wrapping the feet around the studs for steadiness (almost like riding a horse bareback), and moving along as the nailing progresses. Use 16d box nails and place them at least 2 feet o.c., staggering them and keeping them about 3/4 inch in from the edges.

**Cutting In Trimmer Studs.** Trimmer studs, which should be straight and free from serious defects, are required under each end of the top header for every door and window opening. Measure the length accurately and cut each piece slightly "long" to make a tight fit. Spike to the regular studs, using 16d box nails about 18 inches o.c.

**Framing In Lower Headers and Cripples.** The story pole must be checked to determine the exact location of the bottom headers, which are required only in the window openings. Because windows vary greatly in height, the bottom headers will likewise be placed at varying heights.

The length of the bottom header is measured on the bottom plate from trimmer stud to trimmer stud. Each piece is cut separately as measured. Bottom-cripple studs can be cut in the cutting box. Allow as many cripple studs per opening as there are stud marks on the bottom plate, plus two for each opening. These extra studs are placed at the ends of each bottom header.

The bottom header is marked to duplicate the bottom-plate markings. Cripple studs are spiked through the header; the unit is then placed in the wall; cripple studs are toenailed to the bottom plate, and the ends of the header are toenailed to the trimmers.

*Fitting and Nailing Permanent Bracing.* If the opening permits, two braces are required at every outside corner and additional ones are required for every 25 linear feet of wall. Local ordinances should be checked on this point. The ideal slope is a 45-degree angle, but often this angle cannot be obtained because of openings in the wall. It is advisable to make two braces in the same partition oppose each other; that is, do not have the slant in one direction only.

BRACES BETWEEN STUDS. The braces between studs are made of material of the same size as the studs (2 by 4 studs require 2 by 4 braces; 2 by 6 studs require 2 by 6 braces). Each full-length brace is constructed by cutting a series of small braces that fit between the studs.

The top and bottom cuts are made *with a double angle,* so that the center point fits squarely in the corner formed by stud and plate (see Figure 47).

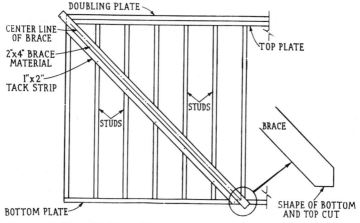

**Figure 47.** Top and bottom cuts of a 2 by 4 wall brace.

It may appear that it is more economical to use scrap pieces of lumber for these braces. Actually, because of the long angle cuts required at the end of each piece of bracing, it is cheaper to use selected "soft" pieces of lumber of sufficient length to go from the top plate to the bottom plate, since one saw cut will make two brace ends. "Tack" a guide piece of 1 by 2 onto the face of the wall or partition, the top edge representing the bottom edge of the brace. Place a 2 by 4 flatwise on the 1 by 2 and mark two vertical lines where it crosses a space between two studs. Then slide the piece of 2 by 4 so that the vertical cut will serve as one end of the next piece of bracing and mark for the other end of the piece.

Continue this marking and sliding process until the piece is marked to make one continuous row of bracing (see Figure 48). Square a line across the face of the lumber and carefully cut on the lines.

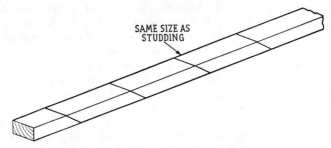

SAME SIZE AS
STUDDING

**Figure 48.** Marking a cut-in wall brace.

Prior to nailing in the brace pieces, start two 8d box nails at each end of each piece. Nails are driven in at right angles to the plumb cut.

When nailing in a series of brace pieces, do not drive the nails in tight until all pieces are in position. This will avoid crowding a stud sideways and will give the completed brace a straight-line appearance.

CONTINUOUS 1 BY 6 BRACE. The second method of bracing is to notch the studs to receive a 1 by 6 continuous brace which runs from the top side of the doubling plate to the lower side of the bottom plate.

On the exterior walls the brace is let into the exterior face of the studs; on inside partitions it makes no difference which partition face is used.

Use 1 by 6 S1S1E lumber to assure accurate thickness and width dimensions, permitting all layout for the "dapping in" to be the same.

**Note:** Building ordinances should be checked, since occasionally 1 by 4 material is permitted. The angle formed by the intersection of a stud with the lower plate is the center of the brace. A 1 by 2 strip (as described above) is tacked onto the face of the studs so that its top edge represents the lower edge of the 1 by 6 brace. Be sure that each stud is straight when the guide strip is nailed on. Lay the brace material on this strip and tack to prevent slipping. Then carefully mark each stud and both plates along the top edge of the 1 by 6. These marks and the top edge of the 1 by 2 indicate where to cut.

The depth of the cut is the same as the thickness of the 1 by 6, usually ¾ inch. Set the Starret-square tongue to ¾ inch and mark both faces of each stud to provide a guideline when chiseling out each dap (see Figure 49).

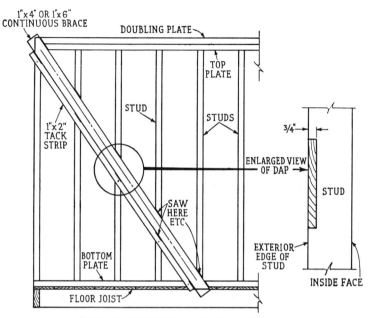

**Figure 49.** Marking for a notched-in wall brace.

Saw all angle cuts and chisel the cuts. Chiseling should be done in reference to the grain, for some grains lead the chisel "in," resulting in a dap that will be rough and too deep. Hence, make the first chisel cut carefully to see from which face of the stud the chiseling should be done.

The top and bottom ends of the 1 by 6 are cut to fit the building. At the bottom-plate line a single angle cut is made; if the top end is started at the outside corner of the building, a combination cut is made.

Two 8d box nails are used in every stud and at both plate intersections.

SUMMARY OF BRACE FITTING AND NAILING:

1. Determine the location of all braces.

2. Determine the kind of bracing required, that is, continuous and dapped into the studs or cut between each stud space.

3. Tack on a 1 by 2 guide strip, on which the brace stock can be rested while marking.

4. Select the brace material, using good-grained "soft" lumber wherever possible.

5. Place a piece of brace lumber in position and mark.

6. Cut the brace into several pieces, if it is the same size as the studs, or notch the studs to the required depth.

7. Nail the brace into place, being sure that each stud is kept in a straight vertical position.

*Fire Stops.* Fire stops are pieces of framing material, the same size as the studding, which are cut and nailed in horizontally halfway between the bottom and top plates.

Their purpose is to shorten the space between the studs to prevent an updraft of air in case of fire.

Note: Originally buildings were framed so that each stud space was open to the attic. Ceiling joists were supported on a 1 by 4 ribbon notched into the inside face of the studs. When this type of framing was used, any fire between the studs at the bottom plate was sucked into the attic because of the immediate upsurgence of air. The horizontal piece of framing stopped this draft— hence the name fire stop.

A second purpose, rarely stressed, is to give additional rigidity to a side wall. A set of studs, after fire stopping, acts as a complete unit instead of as individual studs, thereby increasing the rigidity of the wall.

Fire stops are made from any suitable waste ends that can be found on the job. They must be the same size as the studding. Because of discrepancies in lumber thickness and because studs may not be nailed exactly to their place as marked on the top and bottom plates, it is best to cut each fire stop to the exact length measured between each stud space.

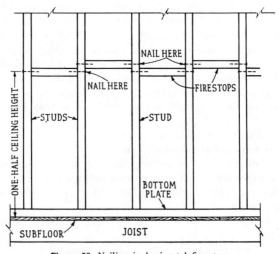

**Figure 50.** Nailing in horizontal fire stops.

Every space must be stopped irrespective of width. In some cases it is necessary to fit them around the wall braces.

Fire stops are nailed by driving two 16d box nails through the studs and into each end of each piece. This can be done if the stops are alternated for height, so that the top line of one piece is the lower line of the one next to it (see Figure 50).

Snapping a chalk line across the face of the studs at the correct height (halfway between floor and ceiling) will facilitate keeping the fire stops in a straight and level line. The final appearance of fire-stop installation, simple as it may appear to be, is another bit of evidence which indicates the craftsmanship of the carpenter.

**Note:** If the exterior walls are finished with vertical boards, additional nailing pieces must be cut in to provide a nailing surface for the boards. These girts are placed approximately 2 feet o.c. (see Chapter 7, page 186). The fire stop will serve as one of the headers.

**Wall Backing.** *Backing* is the term applied to various blocks of wood which must be framed into the walls and partitions of a house to make possible the rigid fastening of curtain rods, wall cabinets, and wall plumbing fixtures. Unless backing is nailed in at exact locations *during the framing process,* it will not be possible to do a first-class job of installing these items after the plastering is completed. Backing has grown even more important since the use of plaster lath has become so widespread; heretofore wood lath could be counted on to hold the drapery-rod screws (see Figure 51).

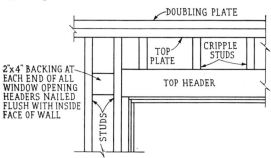

**Figure 51.** Drapery-rod backing.

**Note:** The Molly bolt, illustrated and described in Chapter 9, page 350, is a very excellent device for securely holding drapery rods. It will not do for heavy items, such as a wall lavatory.

The *location* of the backing is very important. The carpenter must ascertain from the plans or from the owner the type and style of drapery rods to be used and plan accordingly. The exposed end of a wall cabinet may come between two studs; hence suitable framing material must be framed in at this location, which can be determined from the detailed drawings of the cabinet (see Figure 52).

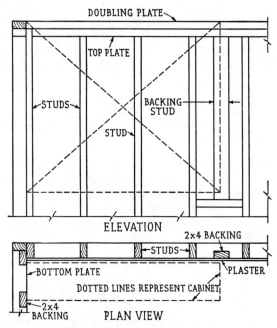

**Figure 52.** Wall backing for end of cabinet.

**Note:** If drapery rods, towel bars, toilet-tissue holders, towel hooks, etc., are to be fastened to the wall with the screws which are usually packed with this type of bathroom wall fitting, then it is most important for the owner *during the framing process* to decide the exact location of these various items and cut in the necessary backing blocks.

After the backing blocks are nailed into place, make a sketch of each wall (labeling it north, south, east, or west) showing the location of the blocks. Each sketch should show floor-to-backing-center measurements, and horizontal measurements from a given wall or partition to each end of the backing. The sketches should be carefully preserved until it is time to install the fittings to the wall.

Too much emphasis cannot be placed on this suggestion; unless it is followed, holes will be bored into finished and painted plaster walls where there is no backing, and hence will be useless. As noted above, the Molly bolt (or equivalent) provides the only alternative way of fastening towel bars and such firmly to the wall in their desired locations.

If partition posts, as described on page 115, have not been used, backing is also required at every inside angle of each room and closet to provide lath nailing. Use a 1 by 6 for a 2 by 4 partition; a 2 by 6 partition will require two 1 by 4s or one 1 by 8 (see Figure 53). Use 8d box nails, toenailed as shown, to increase their holding power. Nail about 16 inches o.c. (Partition studs are recommended.)

Backing for plumbing fixtures, water faucets, etc., is always cut and nailed in by the plumber; the carpenter's responsibility, just prior to plastering, is to be sure that this backing has not been overlooked.

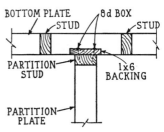

**Figure 53.** Partition backing for inside angles.

Material for wall backing can be made from scrap lumber and waste ends. For drapery rods a piece of 1 by 6, cut in flush with the inside face of the wall, is sufficient. For cabinet ends a piece of 2 by 4, with its flat surface flush with the wall, is best. The center of the 2 by 4 should be placed to correspond with the outside face of the cabinet.

**Note:** On residential repair work the new interior finish should match the old work. If the interior-door casings are 5 or 6 inches wide (this was common practice in house construction of the early twentieth century), baseboard backing is required, as illustrated in Figure 54, to provide nailing for the end of the baseboard where it fits against the door casing. A piece of 2 by 4, 6 or 8 inches long, is sufficient. Each piece of backing should be toenailed with four 8d box nails.

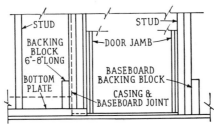

**Figure 54.** Baseboard backing.

*Laying Storm Wall Sheathing.* Storm sheathing is used on the exterior walls of many houses, particularly those buildings constructed in cold climates. The sheathing adds rigidity to the building frame and, by increasing the thickness of the wall, helps to make the house warmer. A heavy paper used between the sheathing and the finish siding gives much added insulation.

Storm sheathing is laid horizontally or diagonally; either method is practical, but laying it diagonally will give an added bracing effect to the frame (see Figure 55).

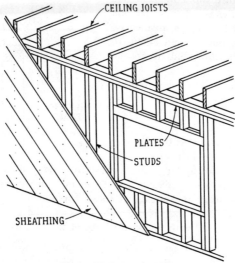

**Figure 55.** Storm sheathing

The material used is usually 1 by 6 S1S1E to ensure uniform thickness and width. The surface side can be laid either in or out; as in subfloor laying, however, the lumber will warp less easily when nailed on with the smooth side next to the studs.

Two 8d box nails are nailed into each stud. Joints are scattered so as not to have too many close together.

When laying the sheathing, start at the sill line and keep even with the bottom edge of the sill to ensure a straight job. Snapping a chalk line to represent the top edge of a first board will result in a straight job with a minimum of effort. A chalk line should be snapped at least every five boards to ensure a craftsmanlike job.

Note: Jobs of this sort, even though they require only rough framing lumber, should be done carefully as they are another bit of "evidence" of craftsmanship. Even a glance at such work reveals the framing ability of the carpenter. It actually costs less to do a true job than to try to correct a job that is neither level nor plumb.

To lay storm sheathing diagonally, start exactly the same way as you would to lay subfloor diagonally. Measure the same distance horizontally and vertically from the lower corner of the building to establish a 45-

degree angle. Strike a line from these two points and then cut, fit, and nail the first sheathing board to this line. Continue this procedure, occasionally striking a line to be sure the front edge of the sheathing is straight. All end joints are cut to a 45-degree angle to make them parallel to the edges of the studs.

*Wall Sheathing for Cedar Shakes.* In many houses a beautiful architectural effect is gained by finishing the exterior walls with cedar shakes, either entirely or in combination with siding. If the house is solid-sheathed, then the walls are ready to receive the shakes; if the walls are not sheathed solid, then it becomes necessary to nail on 1 by 4 sheathing strips, equally spaced, to suit the shake exposure.

Because the layout problems involved in sheathing spacing are so closely related to the operations involved in applying the shakes, the entire process is discussed in Chapter 7, Exterior Finish, page 190.

## PLASTER GROUNDS, USE AND INSTALLATION

At no time can it be assumed that every stud is perfectly plumb, for framing lumber cannot be guaranteed to be perfectly straight, and in spite of careful work, a stud could be slightly out of plumb. Because of these unavoidable discrepancies in framing, plasterers require a guide, called a plaster ground, to make the finished plaster surface true, straight, plumb, and as thick as required by the building ordinance (see Figure 56).

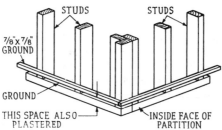

**Figure 56.** Plaster grounds.

Plaster grounds should *not* be nailed in place until the job is ready for the plasterer—all exterior-finish work done, window frames set, the roof shingled, plumbing and wiring roughed in, etc. This procedure will eliminate any chances of grounds' becoming broken or pushed out of line. Grounds are discussed in this chapter since they are used in the framing procedure.

It should also be noted that baseboard is sometimes nailed to place prior to plastering, in which case it serves as a plaster ground (see page 288 for information on the fitting and nailing of baseboard). When doorjambs are set prior to plastering, they too act as a plaster ground around each door opening (see page 276). Window frames are usually set before plastering; hence no plaster grounds are necessary around each window opening.

Grounds are made from strips of lumber cut to an exact dimension to represent the thickness of the finish plaster (including lath). Metal grounds can also be purchased and are particularly useful and valuable on outside corners and archways, for they not only act as plaster grounds but also protect finished corners from becoming damaged.

Plaster grounds are nailed on at various locations on the interior walls and partitions of a building. In well-constructed homes, they are placed at the baseboard line (about 1 inch below the top finish edge of a piece of baseboard) around inside-door openings (unless the doorjambs are set before plastering); on paneled walls where the plastered surface and the paneling are to meet; at the lower edge of a wood cove molding; and at the picture-molding line.

Grounds must be placed so that they are later concealed and, when nailed to place, must be tested for straightness by means of a straightedge. On a residence in which costs are not a major factor, the picture-molding grounds and the baseboard grounds are plumbed to ensure a perfect wall surface.

*Baseboard Grounds.* To install plaster grounds for baseboard, proceed as follows:

1. Order a sufficient amount of ground stock to do the job. To illustrate, the perimeter measurement of each room or closet minus the width of each door opening represents the number of linear feet of baseboard grounds required for that room or closet. Read the specifications or the building ordinance to determine the size of the grounds, usually $\frac{7}{8}$ by $\frac{7}{8}$ inch. Grounds are generally S1E to provide a smooth working surface.

2. Cut lengths to fit each wall space.

3. Determine the baseboard width and cut a small block of wood this length minus the thickness of the ground and minus 1 inch more, to serve as a quick measuring device when nailing on the grounds. This block will make the grounds 1 inch below the top of the baseboard.

4. Rest one end of the ground on this block and nail to the stud, but do not drive the nail all the way in.

5. Continue to nail the ground to each stud in the same manner.

6. Test the face of the ground with a straightedge. If the ground is

"high," remove the nail from the offending stud, chisel behind the ground, removing a little of the stud, and then renail. If the ground is "low," tap lightly until the low spot touches the straightedge, wedge to place, and renail. Obviously high spots must be corrected first.

7. Continue until all grounds are in place.

***Cabinet Grounds.*** Grounds against which a cabinet is to be placed should be carefully plumbed, for the work of installing cabinets is much easier if the plastered surface against which they are to be constructed or fitted is plumb as well as straight.

***Door Grounds.*** Door grounds are installed in the same way as baseboard grounds. If sufficient care was used in selecting straight door trimmers and if the plumbing and straightening of the walls and partitions were accurately done, it is necessary only to nail on the ground around both faces of the opening. The ground should be tested for straightness, and irregularities corrected; that is, high spots planed down and low spots wedged up. The measurement from one ground to the other is the net width of the finish doorjamb stock.

Four times the height of the door opening plus twice its width equals the number of linear feet of grounds required for one door opening.

## CEILING UNIT

***Cutting and Setting Ceiling Joists.*** Ceiling joists are the framing members of the ceilings of a building. There are a number of important items that must be discussed, including (1) the size and length of the material, (2) the direction of the joists, (3) the location of each joist, (4) how to fasten joists to the top plates, (5) how to stiffen them, (6) ceiling backing, and (7) how to frame in an attic crawl hole.

1. The size of ceiling joists is based on the span to be covered. Building-ordinance requirements indicate the required dimensions of the lumber for spans of different widths.

For a room of average width 2 by 4 joists are generally used; a larger room, such as a living room, requires 2 by 6 joists. If the house is two stories high, the floor joists for the second floor act as the ceiling framework for the first story. Second-story ceiling joists are similar to one-story ceiling joists in size, spacing, stiffening, etc.

The length of ceiling joists is determined by the dimensions of the room which they span. (Ceiling joists rarely span from exterior wall to exterior wall, except in small buildings.) Eight inches must be added to the between-the-walls measurement of the room to provide for a joist bearing. This figure must then be increased to an even-foot length of lumber.

A lap joint is used where the ends of two ceiling joists join on a partition. Waste ends are usually not cut off unless they are long enough to make a piece of fire stop (approximately 14 inches).

2. The direction in which ceiling joists are laid is based on a very important construction principle: *the shorter the span, the stiffer the joist.* However, ceiling joists not only provide nailing for the plaster lath but, *of equal importance,* serve to keep the building from spreading. *Hence the shape of the roof must be carefully studied* and ceiling joists placed accordingly. For some buildings ceiling joists may run in two directions, the one at right angles to the other (see Figure 57). This may result in the joists running the long way of a room. Whenever possible, however, the short dimension is preferable.

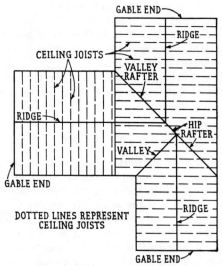

Figure 57. Direction of ceiling joists.

3. Joists are spaced 16 inches o.c. If the stud layout has been done correctly, it is possible to place ceiling joists over each regular stud. No additional layout work need be done.

To retain the correct 16-inch spacing when two joists are lapped, one set is placed directly over the studs, and the second set placed next to the first set and kept parallel to the cross partitions.

4. Joists are placed crown side up (a standard construction principle) and toenailed with two 16d box nails into each bearing.

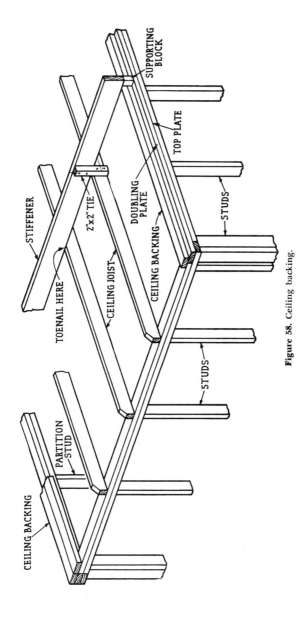

**Figure 58.** Ceiling backing.

5. Wide ceiling-joist spans are stiffened by means of a strongback or stiffener. This piece is placed in the center of the span at right angles to and on top of the joists. The size of the strongback depends on its length. It will vary from 2 by 6 to 2 by 10. If a very wide stiffener is required (because of its length), the joists should be increased in width to secure the maximum rigidity for the ceiling without too much dependence on the stiffener. The ends of the stiffener must rest on a wall or partition. The crown side is placed up, as usual. Two 10d box nails should be toenailed from the stiffener into each ceiling joist. Sometimes a piece of 2 by 2 is used in each angle, permitting spiking into both the ceiling joist and the stiffener (see Figure 58).

*Ceiling Backing.* Ceiling backing is required on top of every partition of exterior wall that runs parallel to the ceiling joists. The material should be 2 inches thick; for a 2 by 4 partition two pieces of 2 by 4 are nailed flat on top of the top plate to give a 2-inch nailing surface for the lath (see Figure 58).

After the backing is on, several 2 by 4 headers should be cut between the two ceiling joists adjacent to each partition to stiffen them (see Figure 58). The backing is spiked to the top plate with 16d box nails spaced at least 16 inches o.c. Nails should be placed as near to the outside edges of the backing as possible to be sure that the piece will not move when the laths are nailed on its underside.

*Framing an Attic Crawl Hole.* An attic crawl hole is often framed in

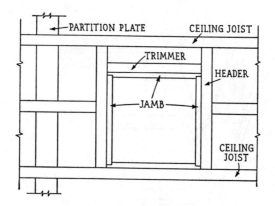

## PLAN VIEW
**Figure 59.** Framing an attic crawl hole.

the ceiling of a closet, where it will be inconspicuous. In a flat-roofed house it should be located where the roof space is the highest. The size of the hole should be not less than 24 by 24 inches.

To frame a crawl hole, proceed as follows:

1. Determine the most advantageous location for the hole.

2. Cut and remove a piece of ceiling joist 30 inches in length, of which 24 inches is for the finished dimension of the hole, 4 inches for two single headers, and 2 inches for the finished-jamb stock.

3. Cut and nail in two headers to form the sides of the hole.

4. Cut in one (or two) trimmer joists to form the other two sides of the hole (see Figure 59).

## R O O F - L A Y O U T   U N I T

The problems involved in roof layout make it the most complicated of all the framing jobs connected with house construction, though stair layout is equally difficult. However, layout work for ordinary rafters for equal-pitch roofs (one pitch only), including common, hip, valley, and jack rafters, can be rather simply described, provided that no attempt is made to discuss the underlying principles of roof framing. Hence, the following pages are limited to only the simplest procedures, which are described on a step-by-step basis to assist the person who desires to lay out, cut, and erect an ordinary roof.*

*Roof Shapes.* Roofs are classified into four main divisions, as illustrated in Figure 60: the shed roof has a slant in one direction; the gable roof has a slant in two directions (it is actually two shed roofs joined together); the hip roof has a slant in four directions; the intersecting roof, which is necessary for buildings with offsets and wings, may be a combination of shed, gable, and hip roofs, requiring a valley rafter where the two different parts of the intersecting roof meet or are joined together.

*Roof Plan.* A plan view of an intersecting roof (looking down from the top) is illustrated in Figure 61. *A plan view of a roof is always the same, irrespective of the slope or pitch of the roof.* This "common denominator" of all roof-layout problems is the basis for simplifying the de-

---

* For persons who desire to pursue the subject of roof framing in its fullest details, including the principles of roof framing, how to solve the unequal-pitch rafter problems, how to use the steel square on various special layout jobs, and how to read the tables given on a good steel square, see J. Douglas Wilson and S. O. Werner, "Simplified Roof Framing," McGraw-Hill Book Company, Inc., New York, 1948.

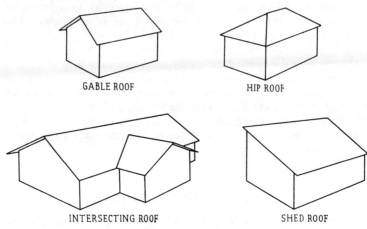

GABLE ROOF

HIP ROOF

INTERSECTING ROOF

SHED ROOF

**Figure 60.** Roof types.

scription of rafter layout. The length of each line is called the *rafter run,* a mathematical term constantly used in rafter layout.

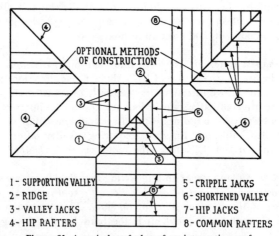

OPTIONAL METHODS OF CONSTRUCTION

1 - SUPPORTING VALLEY
2 - RIDGE
3 - VALLEY JACKS
4 - HIP RAFTERS

5 - CRIPPLE JACKS
6 - SHORTENED VALLEY
7 - HIP JACKS
8 - COMMON RAFTERS

**Figure 61.** A typical roof plan of an intersecting roof.

*Rafters.* Figure 62 illustrates the various rafters required to construct the gable, hip, and intersecting roof, including common rafter, hip rafter, valley rafter, ridge piece, and several kinds of jack rafters. Note that the supporting valley runs from plate to ridge; the shortened valley is framed to the supporting valley. The *top plate* is the horizontal member of a framed wall on which the rafters are nailed.

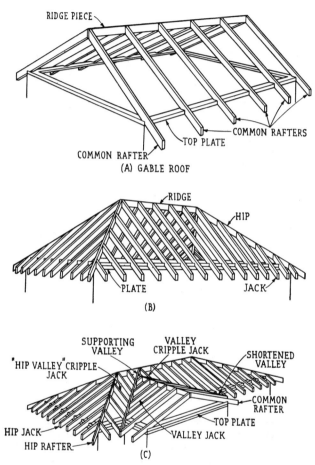

Figure 62. Typical rafters.

*Roof Mathematical Terms.* A person framing a roof should be familiar with a few mathematical terms, as shown in Figure 63.

The *total run* is the horizontal distance over which the rafter passes. For a common rafter, this is always one-half the width of the building. The unit of run, 12 inches, is the basis for all rafter layout.

The *span* is the full-width measurement of the building. The unit of span is twice the unit of run, or 24 inches.

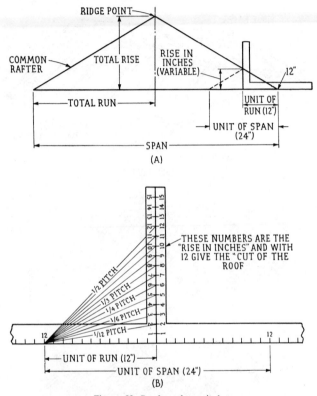

**Figure 63.** Roof mathematical terms.

*Total rise* is the vertical mathematical distance the ridge is above the plate line. For a given pitch and span, this mathematical distance is always the same, irrespective of *the width of the rafter stock*. This is an important fact to remember; a 2 by 4 rafter or a 2 by 10 rafter, mathematically, are the same length for a building that has the same span and roof pitch.

*Pitch* is the mathematical relationship between the total rise and the span. If, for example, a roof has a one-third pitch, the ridge, mathematically, is one-third the width of the span above the plate. To illustrate, a roof span is 30 feet; the roof pitch is one-third; the total rise is 10 feet. For a building 36 feet wide (span) with a one-fourth-pitch roof, the ridge is 9 feet above the plate line. Pitch times span equals total rise. Total rise ÷ total span equals roof pitch. If no pitch is given, measure the total rise on the blueprint; read the span; then apply the mathematical formula and determine the pitch.

In Figure 63B, note the figures on the steel square that give various pitches. The numbers in a vertical direction represent rise; the unit of run (12 inches) represents run. To illustrate, 12 inches and 12 inches gives a one-half-pitch roof; 4 inches and 12 inches gives a one-sixth pitch. Note that the pitch is determined on the basis of the unit of span, 24 inches. Four inches is one-sixth of 24 inches; hence the pitch is one-sixth.

*Common-rafter Terms.* Figure 64 illustrates the following common-rafter terms:

1. *Mathematical length.* The length of a line which has no thickness or width (sometimes known as *line length*).

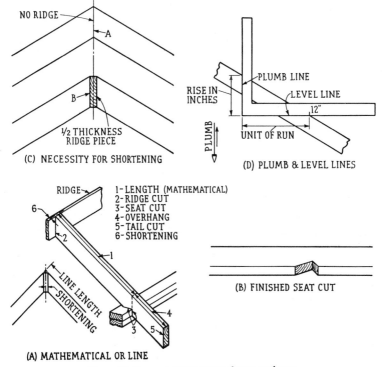

Figure 64. Common-rafter terms, shapes, and cuts.

2. *Ridge cut.* The cut at the top end of the rafter.

3. *Seat cut.* The cut which permits the rafter to fit to the plate. (In trade language, is sometimes known as a *bird's mouth.*)

4. *Overhang.* The part of a rafter which overhangs or extends beyond the roof and forms the frame for the cornice.

5. *Tail cut.* The cut at the lower end of the rafter. This cut often varies in shape to suit the architectural requirements of a house (the tail cut shown in the illustration is a plumb cut).

6. *Shortening.* The layout procedure of shortening the rafter an amount equal to one-half the thickness of the ridge.

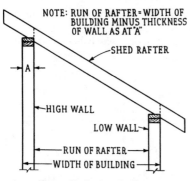

NOTE: RUN OF RAFTER = WIDTH OF BUILDING MINUS THICKNESS OF WALL AS AT "A"

SHED RAFTER

A

HIGH WALL

LOW WALL

RUN OF RAFTER

WIDTH OF BUILDING

**Figure 65.** Shed-roof rafter.

The steel square (with plumb bob) illustrates the meaning of the terms *plumb cut,* a vertical cut after the rafter is in position, and *level cut,* a cut made at right angles to the plumb cut. A *seat cut* is a combination of plumb and level cuts.

***Shed-rafter Shape.*** The shed rafter may appear to be similar to a gable rafter; actually it has two seat cuts, a tail cut, and a top end cut, as shown in Figure 65. The shed-rafter run is slightly different from the common-rafter run in that it is the width of the building *minus* the thickness of the long wall (designated as *A* in the illustration).

***Hip- and Valley-rafter Cuts.*** Hip- and valley-rafter seat cuts are shown in Figure 66*A* and *B*. When top ends of these rafters are fitted to a ridge a single cheek cut is required (see Figure 66*C*). If a hip rafter is fitted to an angle formed by two common rafters (see Figure 72), a double cheek cut is required (see Figure 66*C*). The seat cut for the hip and valley rafters is laid out in exactly the same manner; a close study of the illustration given in Figure 66 will show that the hip rafter has a hollow V, while the valley rafter is cut to a point. The direction for measuring the cheek-cut lines, however, is different; on the hip rafter the cheek-cut line is measured *toward the ridge cut;* on the valley rafter it is measured *toward the tail cut.* A careful study of the numbered items given in the drawing will show the difference in the layout lines. Line 1 in both illustrations represents rafter-length measuring points; line 2 represents the measuring points for the cheek-cut lines.

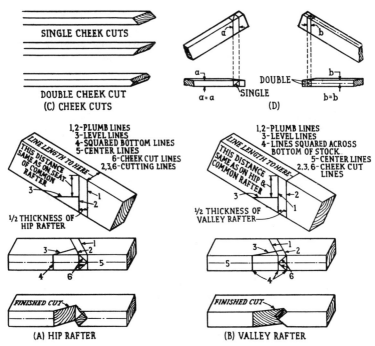

**Figure 66.** Hip- and valley-rafter cuts.

The tail cut of a hip or valley rafter is not shown; for purposes of layout simplicity the tail can be left "long" and cut off after the rafter is in place, the exact cutting points located by means of a straightedge or line at the bottom ends of the common-rafter tails. A true hip-rafter tail cut will be cut to a double miter cut; a true valley-rafter tail cut will be in the form of a hollow V. Often these cuts do not get too much attention since they may be covered with a metal gutter.

**Mathematical Lengths of Rafters.** The length of a rafter is all-important, although the *shape* of the rafter cuts, as previously described, must be carefully laid out if a good framing job is done. If the rafter lengths are not correct, however, the roof will not assemble properly. Figure 67 is the key to the problem of finding the mathematical, or line, length of a rafter.

**Note:** The question of shortening a rafter *to allow for one-half the thickness of the framing stock to which it is framed* is discussed on page 164.

COMMON-RAFTER LENGTHS. The length of a common rafter is the hypotenuse of a triangle formed by one-half the span and the total rise (see Figure 67A). This is determined by square root or preferably by using a steel square to represent the angle (see Figure 67C). Be sure that the square used is stamped (on one face) into twelfths. It is then neces-

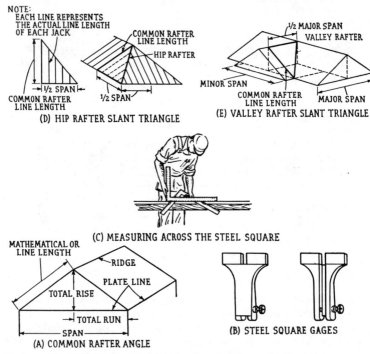

**Figure 67.** Mathematical lengths of common hip and valley rafters.

sary only to measure across from one number on the square which represents the total height to a second number representing the total run (one-half the span) to determine the mathematical length of a common rafter. To illustrate, if the height is 6 feet and the run 8 feet, the diagonal measurement would be 10 feet, the length of the rafter. Again, if the roof has a one-third pitch and a span of 26 feet, the total rise would be one-third of 26 feet, or 8'8" (8 and $\frac{8}{12}$ inches on the square), the total run would be one-half of 26 feet, or 13'0" (13 inches on the square), and the diagonal measurement, from 8'8" to 13'0", would be 15'7½", the length of the rafter. A little practice will permit accurate reading on the square. Square gauges can be purchased to simplify the measuring job (see Figure 67*B*).

HIP- OR VALLEY-RAFTER LENGTHS. The mathematical length of a hip or valley rafter can also be found by means of the steel square. It is necessary to determine the numbers to use for the right angle and measure across this angle to find the hypotenuse of the right angle; this figure is

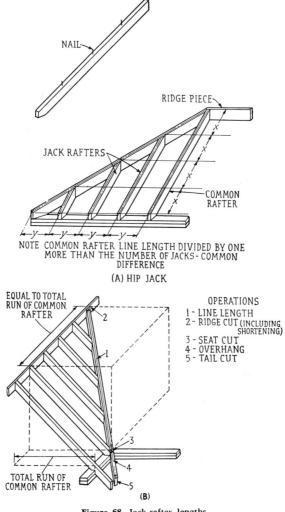

NAIL

RIDGE PIECE

JACK RAFTERS

COMMON RAFTER

NOTE COMMON RAFTER LINE LENGTH DIVIDED BY ONE MORE THAN THE NUMBER OF JACKS = COMMON DIFFERENCE

(A) HIP JACK

EQUAL TO TOTAL RUN OF COMMON RAFTER

OPERATIONS
1 - LINE LENGTH
2 - RIDGE CUT (INCLUDING SHORTENING)
3 - SEAT CUT
4 - OVERHANG
5 - TAIL CUT

TOTAL RUN OF COMMON RAFTER

(B)

**Figure 68.** Jack-rafter lengths.

the length of the hip or valley rafter. Parts $D$ and $E$ of Figure 67 illustrate the angle, known as the "slant triangle" because it is in the slant of the roof. For either a hip or a valley rafter, one leg of the triangle is formed by a common rafter, and the second leg of the triangle is one-half the span of the roof.

To summarize, use the following rules when determining the length of the hip or valley rafter:

*Hip rafter:* Determine the line length of the common rafter (presumably this has already been done).

Divide the span of the roof by 2. Set the square gauges at these two numbers, and measure from gauge to gauge. The result is the line length of the hip rafter.

*Valley rafter:* Repeat the above steps, but be sure to use one-half the major span (main part of the roof) to determine the slant triangle.

**Note:** The slant triangle in $D$ is drawn flat and also shows the line length of each jack rafter. This is explained below.

JACK-RAFTER LENGTHS. (See Figure 68.) A hip jack is actually a shortened common rafter, inasmuch as its seat cut and tail cut are exactly the same as those of the common rafter. The top end requires a single cheek cut (see Figure 66) to make it fit against the hip. The valley jack is likewise a shortened common rafter as its top or ridge cut is made exactly the same as that of the common rafter; its lower end requires a single cheek cut to make it fit against the valley.

All jack rafters have one thing in common: each jack rafter is shorter than a common rafter by an exact mathematical distance, known as the "common difference." This can be determined for the hip jacks by dividing the length of the common rafter by the number of rafter spaces between the common rafter and the corner of the building (see Figure 68$A$). For the valley jacks the same procedure is followed except that the distance to be divided is from the common rafter to the intersection of the ridge cut of the valley rafter with the ridge (see Figure 68$B$).

**Laying Out Rafters by Means of the Steel Square.** (See Figure 69.) Up to this point the discussion has pertained to the *shape* of the rafter cuts and how the lengths of the various rafters are determined mathematically. The terms rise, run, unit of span, unit of run, total run, ridge cut, seat cut, single cheek cut, double cheek cut, ridge, and slant triangle should now be familiar, and a review of Figures 60 to 68, inclusive, should be sufficient to affix these terms in the reader's mind.

COMMON-RAFTER LAYOUT. Using the steel square, a beginner can accomplish the actual layout of a common rafter without too much difficulty. Several distinct steps are necessary.

1. Determine the pitch of the roof. This is all-important.

2. On the basis of the pitch, determine the *cut* of the roof, that is, the rise in inches of the common rafter per 12 inches of run. This is determined by multiplying the unit of span (always 24 inches) by the pitch. The resulting figure is the rise in inches. For instance, a one-third-pitch roof has an 8-inch rise per foot of run; a one-fourth-pitch roof rises 6 inches in 12 inches of run; a one-half-pitch roof rises 12 inches in 12 inches of run (see Figure 63). The cut of the roof is therefore the amount the roof rises (in inches) per 12 inches of run. As illustrated above, the one-third-pitch roof has an 8 and 12 cut; a one-fourth-pitch roof has a 6 and 12 cut; a one-half-pitch roof has a 12 and 12 cut.

3. The cut-of-the-roof figures are used on the steel square to give the correct angle to the rafter cuts. For a one-third-pitch roof set the square

LAYING OUT LINE LENGTH BY LINEAR MEASUREMENT

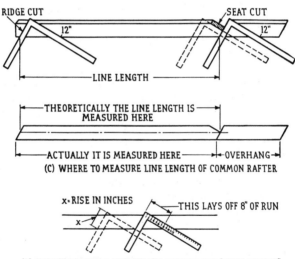

(C) WHERE TO MEASURE LINE LENGTH OF COMMON RAFTER

(B) HOW TO LAY OUT A FRACTIONAL PART OF THE "UNIT OF RUN"

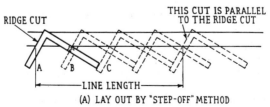

(A) LAY OUT BY "STEP-OFF" METHOD

**Figure 69.** Common-rafter layout with steel square.

gauges at 8 and 12; lay the square on the flat side of the rafter stock, as shown in Figure 69, being sure the crown side of the rafter is up. Marking on the rise figure gives a vertical cut (after the rafter is in position); marking on the 12 side gives a level cut. The seat cut is a combination of both vertical and level cuts.

4. After the square is placed in position, as shown in Figure 69*A*, "step off," that is, move the square as many times as there are feet of total run. Ten feet of run requires 10 steps; 14 feet of run requires 14 steps, etc. Figure 69*B* shows how to lay off a partial foot. In the illustration 8 inches of run is being marked off. Each time that the square is moved, a sharp line is made along the edge of the rafter stock to locate the next position of the square. A check can be made, after the rafter is stepped off, by comparing the length of this line with its mathematical length as determined by measuring the hypotenuse of the correct angle as established on the steel square (see Figure 67*D*). The two figures should coincide if the work has been correctly done; slight discrepancies of ¼ inch may be overlooked.

**Note:** If the length of the rafter has been determined mathematically, the steel square is used as shown in Figure 69*B* and *C* to lay out the ridge, seat, and tail cuts.

HIP- AND VALLEY-RAFTER LAYOUT. The same basic procedure as described above is followed when using the square to step off the length of the hip rafter, with one important exception. The rise per foot of run remains the same as for the common rafter; the run of a hip rafter is always 17 inches (16.97 inches exactly) per 12 inches of common-rafter run (see Figure 70*A*). The reason for using 17 inches is shown in parts *B* and *C* of Figure 70. Briefly, the run of a hip is always the diagonal of a square. A 12- by 12-inch square will always have a hypotenuse of 17 inches. The same rule applies to a valley rafter.

To step off a hip or valley rafter set the gauges at rise in inches and 17 inches and step off as many times as there are feet in the common-rafter run; 10 feet of common-rafter run means 10 steps; 14 feet of common-rafter run means 14 steps, etc. (The same number of steps are taken as noted above for the common rafter, but the steps are larger, that is, 17 inches instead of 12 inches.)

Marking on the rise figure gives a plumb cut; marking on the 17-inch figure gives a level cut.

**Note:** In the layout of the seat cut for a hip rafter, as shown in Figure 66*A*, *be sure to measure the amount of stock left above the seat cut on line 2*. If line 1 is used for this measurement, the top edge of the hip rafter will be too high.

***Cheek-cut Layout.*** The method of laying out a cheek cut requires special consideration. Experienced carpenters use the steel square to de-

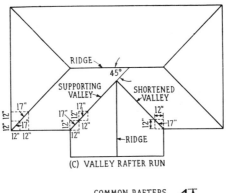

(C) VALLEY RAFTER RUN

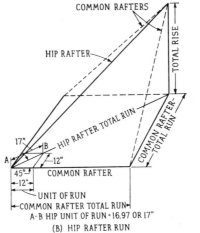

A-B HIP UNIT OF RUN = 16.97 OR 17"

(B) HIP RAFTER RUN

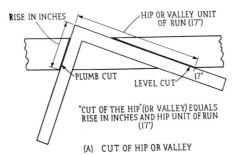

"CUT OF THE HIP"(OR VALLEY) EQUALS
RISE IN INCHES AND HIP UNIT OF RUN
(17")

(A) CUT OF HIP OR VALLEY

**Figure 70.** Angle of cuts for hip and valley rafters.

termine the angle; it can, however, be laid out very simply without this tool. The procedure illustrated in Figure 71 will always give the correct angle, irrespective of roof pitch, as long as the measurement is taken at right angles to the plumb cut.

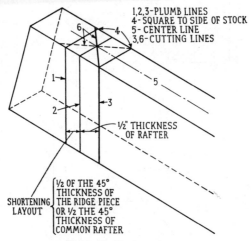

1,2,3-PLUMB LINES
4-SQUARE TO SIDE OF STOCK
5-CENTER LINE
3,6-CUTTING LINES

½" THICKNESS OF RAFTER

SHORTENING LAYOUT

{ ½ OF THE 45° THICKNESS OF THE RIDGE PIECE OR ½ THE 45° THICKNESS OF COMMON RAFTER

**Figure 71.** Cheek-cut layout.

**Note:** Figure 71 also illustrates shortening layout, which is discussed below.

Proceed as follows:

1. Using the square, establish a plumb line on the side of the rafter. For the common rafter and the jack rafter use rise in inches and 12 inches; for the hip and valley rafters use rise in inches and 17 inches. Mark on the rise figure (line 2). Square this line across the top of the stock (line 4).

2. Measure *at right angles* to the plumb line a distance equal to one-half the thickness of the rafter stock (line 3 in the illustration).

3. Mark the center of the stock at line 5, as shown.

4. Connect the top end of line 3 with the intersection of lines 2 and 5. This will give a single cheek cut.

5. For a double cheek cut, repeat the process as shown by the dotted line.

*Shortening.* All common rafters must be shortened an amount equal to one-half the thickness of the stock *to which they are framed.* The reason for this is that mathematical, or line, lengths are, as the term implies, linear measurements only. Figure 64 shows the ridge shorten-

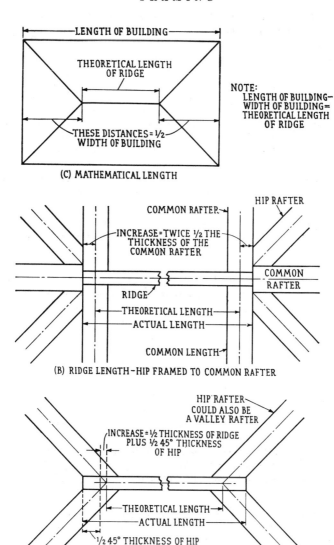

(C) MATHEMATICAL LENGTH

NOTE:
LENGTH OF BUILDING—
WIDTH OF BUILDING=
THEORETICAL LENGTH
OF RIDGE

(B) RIDGE LENGTH-HIP FRAMED TO COMMON RAFTER

(A) RIDGE LENGTH= RIDGE FRAMED TO RIDGE BOARD

**Figure 72.** Ridge lengths.

ing for the common rafter. If the ridge is ¾ inch thick, then the rafter is shortened ⅜ inch. If the ridge is 1¾ inch, then the shortening would be ⅞ inch.

The amount to shorten the hip (or valley) rafter is shown in Figure 72. The only difference from the procedure used for the common rafter is that in this case the shortening is *one-half the 45-degree thickness of the stock* to which the rafter is to be framed.

**Note:** If equal numbers, say 8 and 8, are used on the steel square, the resulting angle on a piece of 2-inch stock will be 45 degrees. One-half the length of the diagonal representing the 45-degree thickness of the stock is then the amount to shorten.

One basic principle of shortening must be emphasized. Always measure *at right angles to the plumb-cut line,* as shown in Figure 71, when allowing for the shortening.

Since jack rafters, either hip or valley, are usually framed to a piece of 2-inch stock, they are shortened one-half the 45-degree thickness of these rafters.

*Ridge Length.* The length of a gable-roof ridge is the same as the length of the plate on which the rafters rest, as shown in Figure 62*A*.

Figure 72 shows, in plan view, the relationship of common rafters and hip rafters to the ridge. In drawing *A* the hip rafter is framed to the ridge; hence a single cheek cut is required. (This situation is also identical for the valley-rafter ridge cut.)

The mathematical length of a ridge for a hip roof is the difference between the width and length measurements of the building, as shown in Figure 72*C*, but the ridge must be made longer than this to permit a full joint between the ridge and the rafter. In Figure 72*A*, note that the length of the ridge is *increased at each end* an amount equal to one-half the 45-degree thickness of the hip rafters.

In Figure 72*B*, the hip rafters are framed in the angle formed by three common rafters (a procedure known as the tripod method), and the ridge is increased at each end by one-half the square (90-degree) thickness of the common rafter.

**Note:** The above description will enable an experienced carpenter to determine the exact length of ridge board required. In actual practice, if desired, the length of the ridge can be estimated, cut a little long, and then cut off *after* the roof is raised. For an inexperienced roof framer, this is probably the safest procedure to follow.

## RAISING A ROOF

In order to know how many rafters of each kind are required, the carpenter should locate the position of each rafter *before* he cuts the rafters. The layout marks are made on the top plates of the walls that are to support the rafters.

Several important steps are necessary *prior* to raising the roof (see Figure 73). On the top plate of the framed walls which are to support the rafters mark the location of each rafter. Ordinarily, rafters are spaced 24 inches on center; thus every third rafter will be placed against a ceiling joist, to which it can be spiked. However, the appearance of the overhang part of the rafter, if exposed to the eye, must also be studied in order to make the rafters appear equally spaced. For a building that has a gable roof and is exactly an even number of feet long the problem is relatively simple, for the length of the building can be divided by 2 feet (24 inches), with nothing remaining. If a building is, say, 32'8" long, then either the additional 8 inches must be left as a narrow space or the rafters placed closer together to make all spaces of equal width.

There is no hard and fast rule that can be stated on this rafter-spacing layout problem. In a shed building used for farming, obviously, the spacing of the rafters is relatively unimportant. On a costly residence the problem is much more important. Care should be taken to be sure opposite sides of the building are laid out exactly alike, so that each pair of rafters will parallel the walls of the building.

*Ridge-board Layout.* Select a straight piece of lumber for the ridge board, usually 2-inch stock, and lay out the rafter locations to correspond with the rafter marks on the plates. The marks should be squared across both faces (the flat side) of the ridge board to aid in keeping the rafters in a square position (not twisted) when the ridge board is nailed to the rafters or vice versa.

*Scaffolding.* The height of the ridge board above the plate line determines whether a scaffold is needed for raising the roof. If the roof has a steep pitch, then a scaffold, as illustrated in Figure 73, will be required. Sometimes a pair of sawhorses, carefully placed on boards laid at right angles to the ceiling joists, will suffice. The scaffold should always be about 3 feet below the final position of the ridge board.

*Placing the Rafters.* After the rafters are cut and the scaffolding is ready, the rafters are placed in their approximate location. To avoid

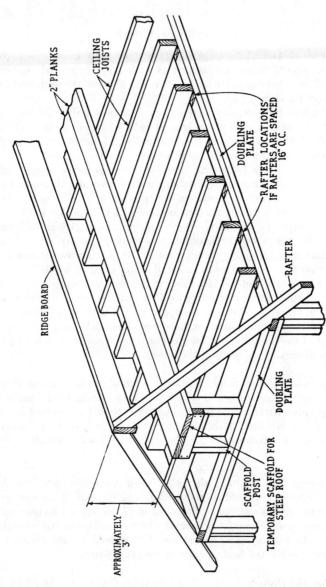

**Figure 73.** Raising a roof by means of a scaffold.

2" PLANKS

CEILING JOISTS

DOUBLING PLATE

RAFTER LOCATIONS IF RAFTERS ARE SPACED 16" O.C.

RIDGE BOARD

RAFTER

DOUBLING PLATE

SCAFFOLD POST

TEMPORARY SCAFFOLD FOR STEEP ROOF

APPROXIMATELY 3'

having to turn the rafters, be sure that the ends cut to fit against the ridge are in the correct position.

*Raising the Rafters.* To raise the rafters, proceed as follows:

1. Select a straight pair of rafters for the ends and nail the ridge board to one of them.

2. Hold the ridge in a level position (this will require someone to hold up the free end), be sure that the rafter seat is tight to the plate, and nail to the plate, using 16d and 8d box nails.

3. Nail the opposite rafter in like manner; this will form the end gable and immediately cause the ridge to be self-supporting at one end.

4. Repeat this procedure at the other end of the ridge board, and nail a second pair of rafters to place. These need not be selected rafters, unless they form a gable end.

5. Temporarily brace the four rafters and ridge by nailing on a brace from ridge to plate, keeping it on a 45-degree angle.

6. Fill in the intermediate rafters, a pair at a time. Nail through the rafter into the plate, and toenail 8d box nails into the ridge board.

7. Continue with the next section of the roof in a similar manner.

Hip and valley rafters need careful placing and nailing to the ridge board and wall plate. They must also be sighted sideways and temporary braces put on to hold them in perfect alignment; otherwise they may be crowded to one side when the jack rafters are nailed to place, and the result will be a crooked hip or valley. Jack rafters are always nailed in opposing pairs.

To be sure that the end of each exposed rafter (excepting the hip or valley rafters) is at a 90-degree angle with the plate line, sight *each* rafter sideways and hold to a straight position by temporarily nailing a sheathing board on the top side of the rafters. This board is placed approximately halfway between the ridge and the bottom plate.

## FINAL STEPS IN FRAMING A ROOF

*Bracing a Roof.* (See Figure 74.) No specific instructions can be given regarding the bracing of a roof except to emphasize a few procedures that will ensure a true roof surface that will not sag.

1. A 2 by 4 or 2 by 6 is often nailed (on edge) under the rafters at about the center between the ridge board and the wall plate. This is sometimes called a *purlin.*

2. The purlin is held up by posts, which should rest on or very near a partition and should be placed at 90 degrees to the purlin. Sometimes the purlin is moved "off center" to permit a better post arrangement.

3. Braces are cut in the horizontal direction of the purlin to prevent end movement.

4. Roof braces should never be placed to permit a roof load to be carried to the room ceiling. When no brace support can be found, a

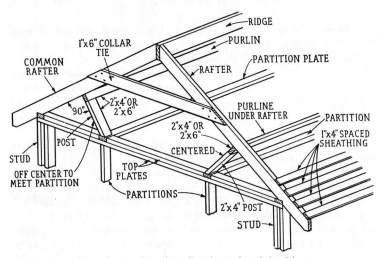

**Figure 74.** Roof purlin, collar ties, and roof sheathing.

timber, on which purlin posts can be fitted and nailed, must be placed over the ceiling joists (at least ¾ inch above them) to prevent the transmission of load.

*Collar Ties.* As shown in Figure 74, collar ties, preferably of 1 by 6 lumber, are used with or without purlins. A tie should be nailed on about every fourth pair of rafters. Long ties need a vertical support to the ridge.

*Gable Studs.* Gable studs, illustrated in Figure 75, should be placed directly over the wall studs, 16 inches o.c.

Gable studs must be carefully measured and nailed, as it is quite easy to get a crown in the rafter. To avoid crowning the rafter, tack on a 1 by 4 flat brace against the outside face of the rafter and the top plate. Sight the rafter for perfect alignment before nailing the brace.

The studs are set on edge and notched around the rafter.

The top ends of gable studs are cut to suit the roof pitch. The angle of the cut, using the steel square, is rise in inches and 12 inches; cut on the rise. To illustrate, a one-fourth-pitch roof has a 6 and 12 cut. Place

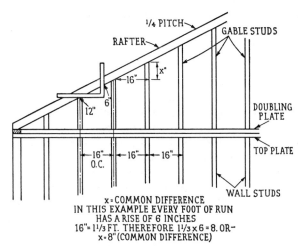

**Figure 75.** Gable studs.

the square on the gable-stud stock to 6 and 12 and mark on the figure 6; this will be the correct angle for cutting the top end.

For 16-inch spacing (1⅓ feet) multiply the rise in inches (6 inches) by 1⅓. The product is 8 inches, the common difference in the length of each gable stud. As soon as the longest stud is cut into place, each succeeding stud will be 8 inches shorter. This common difference varies with the roof pitch.

Gable studs are toenailed into the plate with 8d box nails and spiked and toenailed into the rafter with 16d and 8d box nails.

*Sheathing a Roof.* (See Figure 74.) If a roof is to be covered with composition roofing, it must be sheathed solid. Ordinarily 1 by 6 S1S1E stock is used. If wood shingles are specified, use 1 by 4 stock S1S1E and make the space between boards equal to the width of the sheathing. To simplify the spacing, carpenters use spacing blocks to separate the boards as they are cut and nailed to the roof. Nail each board to each rafter, using two 8d box nails per rafter. Cover the ends of the sheathing boards as described in the Cornice Work section of Chapter 7.

When nearing the ridge with solid sheathing but before reaching the peak, fit and nail the two sheathing boards which form the peak of the roof; then fit the last boards in the spaces that are left. This method is the easiest way of ensuring a straight ridge joint; discrepancies in the width of the last board to be fitted can be corrected by ripping or cutting with a hand ax.

If the rough side of the sheathing boards is exposed, the boards will not warp so easily.

## STAIR UNIT*

Stair-layout work involves considerable mathematical ability to determine correctly the exact rise and run of each stair tread as well as the "over-all" measurements of a stairway, including (1) the total rise, which is the vertical distance from finish floor to the finish floor above, and (2) the total run, which is the horizontal space occupied by the stairway.

Since most new dwellings are only one-story, the problem of stair layout is not a common one. Also, because of the complexity of stairbuilding, which is usually done by a specialist in this field, no description will be given regarding this operation.

* Those who desire to pursue this subject in detail should see J. Douglas Wilson and S. O. Werner, "Simplified Stair Layout," Delmar Publishers, Albany, N.Y., 1947.

# CHAPTER SEVEN

# Exterior Finish

The exterior-finish work involved in completely closing in the frame of a house can be divided into four main divisions, namely, siding, frames, cornice, and roofing.

If a house is to have a plastered exterior, there will be no need for siding, and the work will be done by the plasterer. Many houses built today have little cornice work; flat-roofed houses have none.

Each of the four topics listed above will, however, be described in detail to provide adequate coverage for any type of home that may be constructed or added on to by a prospective homeowner.

The tools required to do the exterior-finish work of a house include hammers, finish saws, chisels, planes, and oil and carborundum stones. The layout tools are the same as used during the framing process: the steel square, Starret or try square, T bevel, rule, level, building twine, and straightedge.

Before any exterior work is started, the cutting tools should be sharpened in readiness to do some fine joinery. Exterior-finish work furnishes a splendid opportunity for an apprentice carpenter to secure practice in making precision joints. Even though some of the joints may be opened up by the hot rays of the sun and all joints will eventually be painted (and puttied if necessary), the carpentry should be first-class: cornices should be straight, siding should be level and straight, and frames should be square and set plumb and level. All joints should be accurately made.

Exterior-finish materials are made from weather-resistant lumber. Redwood is probably the best known lumber used for siding, cornice members, and the exterior casings of window and doorframes. Western red cedar, southern white cedar, and Port Orford cedar are also excellent materials which will hold up under almost any weather condition.

Architecturally, the siding, frames, and cornices, when well designed, make the exterior of a residence beautiful; the size and durability of the various finish members, therefore, are more important than their strength. Frame jambs, however, must be large enough to provide rigidity to the frame and holding power for the hinges.

The plans and specifications should be carefully checked prior to making out the bill for the exterior-finish materials, as architects will always specify the kind and quality of lumber to use for siding, cornices, and frames. Specifications are based on the materials carried in stock by the local lumberyard. The lumber dealer should therefore be consulted if a home builder is proceeding without benefit of specifications covering the exterior-finish items.

Note: Building ordinances rarely, if ever, state lumber requirements for exterior finish, for cornices, frames, and siding are not structural items. They have been designed, not to carry a load, but merely to provide an exterior covering for a building.

Some exterior-finish jobs require a scaffold; this is necessary when doing cornice work and, usually, when applying siding to the upper part of the walls. Sometimes "high" horses are built to eliminate the necessity for constructing a staging; this procedure is best suited to the needs of a contractor who makes continuous use of such equipment.

A good workbench is essential when constructing the window and doorframes. The bench should include a vise or bench block to hold materials securely while they are being worked on.

Note: Many contractors will order ready-made frames from a mill. Obviously, a machine-made frame can be cut with a somewhat greater degree of accuracy than one made by hand. Window and doorframes, however, are still made by hand; a person living in a rural district may be far removed from a mill. Also, since the intent of this book is to explain operations which can be accurately performed by hand, enabling the homeowner to have the satisfaction of doing his own work, it is essential to describe the equipment on which various operations involved in frame making can be performed.

## CONSTRUCTING A SCAFFOLD*

A scaffold for residential construction is composed of uprights or poles; horizontal members, called ledgers, on which the planks rest; ties; braces; and handrails. Safety ordinances are quite specific about the construction requirements pertaining to the maximum span between

* The above description is based on the pamphlet "Your Life May Depend on a Safe Scaffold," published by the Division of Safety, Department of Industrial Relations, State of California.

ledgers, usually not more than 10 feet for "light-trades" scaffolding, and ledger length, which is not to exceed a 3-foot span. (See Figure 1.)

Horizontal ties should be placed so as to act as a support for the ledgers. Planks should be no smaller than 2 by 10s, and each end must have full ledger support; that is, the end of one plank should not rest on an end of the next plank.

Bracing is placed at a suitable angle, from the top of one pole to the bottom of the next one. Poles are usually 2 by 4s and must be free from large knots or other serious lumber defects. Each pole should rest on a suitable block, say 2 by 10 by 10 inches, to prevent it from settling. A continuous shoe is better on soft ground.

**Note:** Certain types of scaffold are prohibited by law: the shore, or single-leg type; the lean-to, a ledger fastened to one pair of legs only, with the inner end of the ledger resting against the house and depending entirely on the pressure against the building to hold it level; and the bracket scaffold (unless it is bolted through the wall or hooked over the top of the wall).

A handrail is placed approximately 3 feet above the planks and is fastened to the inside of the outside set of uprights. Figure 1 illustrates the basic principles of scaffold construction.

The weave is removed from a scaffold if it is tied to the frame of the building, sometimes to an exposed rafter end and sometimes into an opening. (This can be done prior to the setting of the frame.)

To permit work room next to the building, a scaffold should be placed at least 12 inches from the building. The height of a scaffold is quite important. For a high scaffold (used in working on a two-story house, for instance) the ledgers and planks are placed a maximum of 7′6″ apart; this distance is the carpenter's workable wall space, and he can work successfully within this range. If the scaffold is for cornice work only, the ledgers and planks should be located about 3 feet below the roof line.

Ledgers must be carefully nailed. Since they are usually made from 1 by 6 or 1 by 8 stock, three 8d common nails should be used at each ledger end. Double-headed scaffold nails can be bought which are the same length as 8d nails under the second, or lower, head; these nails are easily removable and make possible the recovery of more lumber (and also save more time) when the scaffold is dismounted.

Horizontal ties, handrails, and braces should have two nails per post. A scaffold must *always* be nailed so that each nailed joint is strong enough to support the weight of a man.

To construct a scaffold, proceed as follows:

1. Select uprights of suitable length and free from serious defects. Ordinarily, 2 by 4s are used since they are usable in other parts of a building.

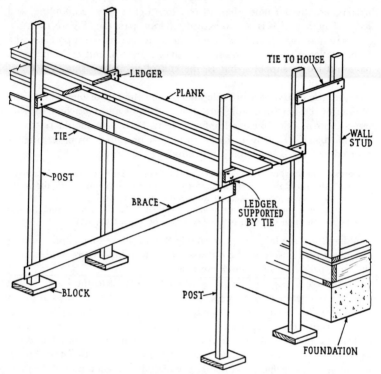

**Figure 1.** Basic principles of scaffold construction.

2. Determine the number of posts and ledgers needed. The length of the wall divided by 10 will give the number needed for one side of a building. They should be equally spaced or conform to the length of plank to be used.

3. Select a flat piece of ground and nail a ledger (or ledgers) to two poles at a predetermined height. Continue until all sets of standards are completed.

**Note:** On side-hill construction it may be necessary to make each ledger height different on each set of poles in order to make the scaffold planks level. In this case determine the scaffold height in relation to the framed building —for instance, 4'0" from the floor height or 3'6" from the top-plate line, etc.

4. Raise each standard to place, temporarily bracing the first set to the building. Nail on the horizontal ties as the raising work continues, being sure that the tie is placed tightly to the underside of each ledger.

5. Diagonal-brace the scaffold to assure rigidity.

6. Place planks on ledgers, being sure that each piece is sound. Each plank should be pretested for soundness by placing it on two pieces of lumber with the center free and "jumping" on it. This procedure will reveal any hidden defects.

7. Nail on handrails.

Note: A simplified type of scaffolding can be constructed by using one post only; the inside end of each ledger is nailed into a window opening or onto one edge of a wall stud. When this type of construction is used, it is customary to complete the siding work above the openings and construct the cornice. The scaffold is then removed, and the remainder of the exterior-wall finish is completed from the ground by means of a pair of carpenter's horses. Obviously, this procedure is possible only on a one-story house constructed on a reasonably level lot (see Figure 2).

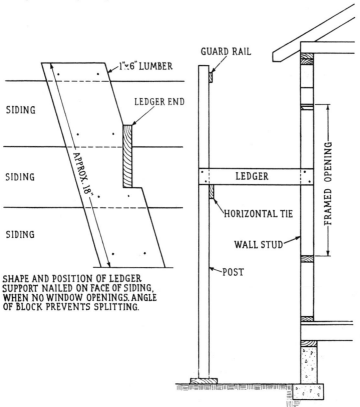

**Figure 2.** Simplified scaffold.

## CONSTRUCTING A CARPENTER'S WORKBENCH

A workbench, for job usage, must be light enough to facilitate moving but strong enough to withstand considerable hammering. It should be approximately 30 inches wide and 8 feet long. The height may vary somewhat according to the height of the person using the bench, but 34 inches is recognized as standard.

To make a workbench, proceed as follows:

1. Select four 2 by 4s (see Figure 3) and cut to 33 inches in length for legs.

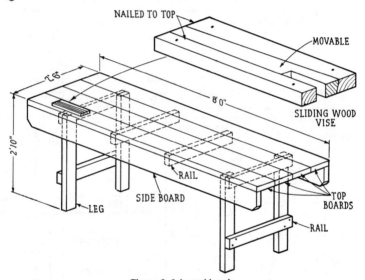

**Figure 3.** Job workbench.

2. Select two 1 by 6s and two 2 by 4s for horizontal members and cut to 30 inches in length.

3. Nail one 2 by 4 to the top of two uprights or legs and one 1 by 6 approximately 12 inches from the bottom of the legs. Repeat this procedure for the other set of legs.

4. Select two suitable pieces of 1 by 12 stock and cut to 8'0" length. These are for the sides of the bench.

5. Nail one sidepiece to the legs, keeping the legs 12 inches in from each end of the piece.

6. Nail on the other 1 by 12.

7. Cut in two crosspieces between the sidepieces to form support for the toppieces.

8. Select 1-inch surfaced stock (1 by 8, 1 by 10, etc.) for the bench top and nail to place. Vertical-grained lumber, if it can be found, is preferable because it will not sliver. Joints between the top boards need not be tight; in fact, a little space between them provides an excellent slot in which to drop a saw when it is not in use. Use 8d box nails (to prevent warping) and set each nail to protect the plane-bit edges as they are laid on the bench.

9. Cut in braces at a 45-degree angle (see illustration) to make the bench rigid.

10. Make a 1-inch sliding wood vise, as shown in Figure 3. Metal vises can also be used by clamping them to the ends of the bench top.

## CUTTING, FITTING, AND NAILING SIDING, GENERAL INFORMATION

The following basic information is not given in any particular construction order as certain items may be first in order of importance on one job while other items may need emphasis on another job. Hence it is important to get an over-all understanding of all phases of siding, fitting, and nailing.

There are two basic methods used to enclose the exterior walls of a framed house:

1. Siding materials are nailed on horizontally. A rabbeted joint at the lower edge of each board provides a reasonably tight joint.

2. Boards are nailed on vertically; this method is known as a board-and-batten finish. In this case a crack is left between each board and later covered with a batten, usually a $\frac{3}{8}$- by $2\frac{1}{2}$-inch piece of S4S lumber. Sometimes a molded batten is used that is $\frac{3}{4}$ by $2\frac{1}{2}$ inches in size.

Generally, vertical siding is used to make a house, or some portion of it look higher, while horizontal siding "stretches out" the house, making it look longer or wider. However, shingle, or shake, siding, because of its distinctive character, is gaining in favor.

On houses that have been storm-sheathed the frames and corner boards are set first, and the siding is then fitted tightly to the edges of the frame casings and the edges of the corner boards. In this case, a strip of building paper 6 inches wide is placed under the casings and corner boards to make a windproof joint when the paper is placed on the wall under the siding.

On houses that have not been sheathed the siding is nailed on first, and then the frames are set so that the casing rests on top of the siding.

This method, which is also used for the corner boards, is not recommended for cold climates.

Siding is made in a variety of shapes and sizes, varying from $\frac{5}{8}$- by 4-inch siding with a net coverage of 3 inches, which has a sawed surface to receive stain, to $\frac{3}{4}$- by 10-inch channel siding with a net coverage of 9 inches, which has a smooth surface to receive paint (see Figure 4). The shape to use is based on the architectural effect desired; differences in effect are also obtained by the use of corner boards in place of miter joints at all corners of the building.

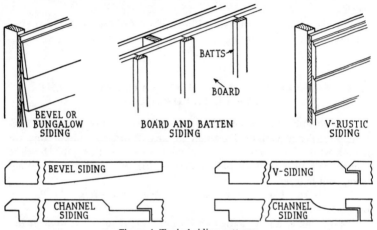

**Figure 4.** Typical siding patterns.

Vertical boards are usually $\frac{3}{4}$ by 10 inches or $\frac{3}{4}$ by 12 inches, S1S1E. The exterior face can be either smooth or rough, depending on whether stain or paint is to be used.

The various kinds of lumber used for exterior-wall finish have been described on page 17.

A heat-resistant building paper is used underneath the siding of the vertical boards to make a more weather-resistant wall, both when nailed to storm sheathing and when nailed directly to the studding. This keeps the house warmer in winter and cooler in summer. If a building plan is being followed, the specifications will indicate what kind to use. There are many kinds available; some have much more insulation value than others. In any case the paper is purchased in 36-inch rolls. The number of square yards in a roll may vary.

Siding should be stored in a clean, dry place until it is applied and the walls should be dry at time of application. Never do the work when it is raining or immediately after a storm. If a house is storm-sheathed and is to be plastered on the interior, postpone applying the siding until most of the excess moisture in the plaster has been driven off, even then making certain of sufficient ventilation through the window openings until the plaster is thoroughly dry.

*Applying Building Paper.* It is customary to nail on building paper a strip at a time. This method keeps most of the studs exposed so that they can be easily located when nailing on the siding. Only a few nails are needed in the top edge of each strip of paper to hold it to place, for the siding nails will hold the paper tightly to the studs or storm sheathing as the work progresses.

Strips of paper are cut the length of the wall. To simplify the work, the paper is cut to length and nailed to place as the siding work proceeds. A pocketknife or linoleum knife is best for cutting the paper.

Allow ample overlap of each layer of the paper where the ends and edges join. Provide a double layer at corners of the building. These precautions ensure greater protection against infiltration of wind and moisture.

If board and battens are used, the building paper is cut the same length as the height of the wall, one strip at a time, and nailed on as the work progresses.

## HORIZONTAL SIDING

*Making and Using a Siding-layout Rod.* It is easy for siding to get out of level during the process of fitting and nailing it to the wall. On large jobs, where several carpenters are employed, this problem increases in difficulty if the craftsmen work at the same time on different elevations of the building. The procedure is greatly simplified if a siding rod is used.

The correctness of the wall layout shows up quickly when a siding rod is laid out since the total height of the wall from the top of the foundation to the top surface of the top plate should have been adjusted, if possible, to conform to an even number of siding boards in order to eliminate the chance of having to use a part of a board at the top.

To make a rod, proceed as follows:

1. Select a piece of 1 by 1 material a little longer than the over-all wall-height dimension from sill to top plate.

2. Measure the exact wall height from foundation to top plate and mark this distance on the rod.

3. Carefully measure several siding boards to determine the average

wall coverage per board. For instance, a 6-inch siding board will ordinarily measure 5 inches; from the top edge of the rabbet to the top edge of the board, it might be 5¼₆ inches.

4. Set a pair of dividers to this dimension.

5. Using the dividers and starting at one end of the rod, step off a series of steps until the top end of the rod is reached.

6. Square a line at each step. The rod now represents the exact location of the top edge of each row of siding.

7. Carefully locate and mark on the building the top edge of the first siding board. This should be done so that the lower edge of the siding covers the joint between the top of the foundation wall and the lower side of the mudsill. Do not forget to allow for the rabbet.

8. Tack the rod in a vertical position to a corner stud, being careful to see that the top edge of the first siding board marked on the rod exactly

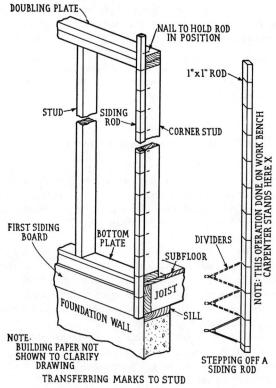

**Figure 5.** Siding-rod layout and usage.

coincides with the top edge of the first siding board as marked on the wall of the house (see Figure 5).

9. Carefully transfer the marks representing the top edge of each siding board to the outside edge of the corner stud.

10. Repeat this process for every corner, every inside angle, and on the two trimmer studs of every major window and door opening.

**Note:** If building paper is used, the layout rod must be used *after* a strip of building paper is nailed on.

Once the siding layout is completed, it is possible to strike chalk lines for the long siding runs, using the siding marks to locate the position of the line. All corners will coincide exactly, provided that each succeeding siding row is laid to the line, which represents its top edge.

**Note:** Because the location of each horizontal board is known, a carpenter can start the siding above the window line and continue to the top-plate line *before* sheathing the roof. This greatly simplifies the fitting and nailing of the siding at the top of the wall. Then the carpenter can start the remainder of the siding from the bottom with full assurance that the last board which meets the siding first put on will exactly fit its place.

*Making a Siding Miter.* To miter a piece of siding for an exterior-corner joint requires considerable skill with a finish saw. A miter block is often used, and the siding is cut from the back side, as shown in Figure 6. This method makes possible the true cutting of a miter, irrespective of the shape of the siding on its finished surface. The angle of the miter cut is 45 degrees "plus," in order to make a "long" miter, one that will be slightly open on the inside (where it cannot be seen) and cause the two pieces of siding to be mitered together to fit tightly.

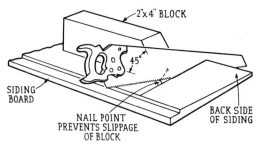

**Figure 6.** Siding miter block.

Butt joints are cut either in a specially prepared wood miter box, made wide enough to receive the siding, or by the ordinary process of squaring a line on the face of the stock, cutting on the line, and then block-planing the end to a smooth cut.

*Fitting Siding to an Inside Angle.* An inside angle presents a new joinery problem when two pieces of siding are to be butted at a 90-degree angle. There are two methods used:

1. Make a scribed joint by marking with a pair of scribers and then cutting on the line to make the end conform to the shape of the siding. This is not difficult on a simple siding design but can be quite complicated if the siding is a molded one.

2. The simplest method, and a practical one, too, is to prepare a square strip ¼ inch thicker than the siding. A ⅝-inch siding requires a ⅞- by ⅞-inch strip; a ¾-inch siding requires a 1- by 1-inch-net strip. This strip is nailed into the inside angle, and the siding is butted to the strip by cutting a butt joint (see Figure 7). The length of the butt-joint strip is from the foundation to the top plate. If a piece of "soft" material is selected, it will not split as it is nailed into the angle.

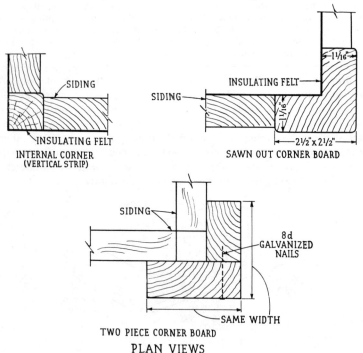

**Figure 7.** Inside-angle strip and corner boards.

*Selecting Siding Lengths.* Siding should be carefully selected for length to eliminate as many butt joints as possible. For instance, a wall 13′6″

long should be covered with 14'0" lengths of siding assuming, of course, that such lengths are available on the job. A wall 29'0" long will require at least two lengths of board, and even then succeeding rows of siding must be started with short lengths if the butt joints are to be scattered.

Boards that are slightly defective or marred on their finish surface should be placed near the top of a wall, where they will be less conspicuous.

*How to Nail Siding.*\* When applying siding over wall sheathing, use 10d nails on the same schedule as noted below for 8d nails. In 10-inch flush siding, drive three 8d common nails per bearing, placing one ¾ inch from each edge (or in the quirk of a molded edge) and the third at the middle of the piece. In narrower flush sidings, use two nails per bearing.

Galvanized siding nails or box nails may be substituted. Ordinarily, uncoated common nails will suffice, since western pines do not contain any objectionable acids which attack nails. Galvanized nails are always better for outside use, though, as they resist rust, which, if allowed to develop, will stain the finished paint coat at the nailheads.

When nailheads are countersunk into the wood which is to be painted, always fill the holes with white-lead putty after the prime coat is applied to the siding.

*Summary.* Considerable tool skill and craftsmanship are required to fit and nail siding to its place. The siding must be started correctly in relation to the top of the foundation; it must be kept perfectly straight and at the same height as the work progresses; butt or end joints must be scattered; that is, two siding boards in adjoining rows should not be joined on the same stud; miter joints should be perfectly made so that they are not open on the face. The nails used must be of the correct size and must be driven carefully so as not to leave hammer marks.

Assuming that a siding rod has been prepared and that all corners, angles, and openings have been marked, proceed as follows when fitting and nailing on siding:

1. Distribute the siding material according to lengths that will eliminate butt joints.

2. Cut and put on the first strip of paper.

3. "Strike" a chalk line representing the top edge of the first board.

4. Nail the first siding board to place, keeping the top edge exactly even with the chalk line.

\* From the pamphlet "Economy Sidings in Western Pine," Western Pine Association, Portland, Oregon. (Used by permission with appreciation by the author.)

5. Continue until the job is completed. Make all joints true, keep the siding level and hold each board to the siding-layout lines.

## BOARDS AND BATTENS

Boards nailed vertically on the exterior walls of a house form what is known as a *board-and-batten finish*. A layout rod is not necessary for this type of wall finish. The boards should be started from the center of a wall and laid in opposite directions until the ends are reached; this procedure will assure having the same-width board at each end, a requirement of good design. It is advantageous first to determine the total wall length and divide this measurement by the width of one board, then to spread the boards apart a sufficient amount to make sure that a narrow strip of board will not be required at each end of the wall. The spaces must, however, be less than the width of a batten to provide good nailing for the battens. Horizontal headers 2 feet o.c. provide nailing for the boards.

Boards are cut to length from slightly below the top of the foundation wall (to make sure that this joint is covered) to the top of the plate line. Cutting around the window openings is done as needed.

Battens are usually *not* cut and nailed on until after the vertical boards have had a chance to finish shrinking. The exposure to the sun of each board, dry as it may already be, will cause even the best air-dried lumber to shrink a bit more. Sometimes battens are cut, fitted, and nailed to place, but *nailed only on one edge;* the second edge is nailed as almost the last job on the exterior finish. This method of nailing permits board shrinkage and prevents the battens from splitting, as they might do if nailed immediately to two separate boards.

Vertical boards are either 1 by 10 or 1 by 12 stock. They should be nailed to each header or girt and each top and bottom plate. Nails should be kept $\frac{3}{4}$ inch in from each edge, and one nail should be driven through the center of each board into the header. It is best to use 8d galvanized box nails as they provide a shank long enough to enter the header and their flatheads hold the boards from warping.

Battens are nailed at least every 12 inches; the size of the nail depends on the thickness of the batts; 5d or 6d cement or galvanized nails are excellent because of their slimness, length, and holding power.

*Summary.* To secure a first-class job of fitting and nailing vertical boards and battens, proceed as follows, assuming that all necessary nailing girts or headers are framed in:

1. Determine the width of the boards by reading the plans or specifications or by measuring the material as delivered to the job.

2. Locate the center of *each* exterior wall.

3. Compute mathematically the required spacing between each board. This must be done for each wall.

4. Lay off on each wall the location of each board.

5. Select any given wall and nail on the first center board. Be sure that this and all subsequent boards are plumbed. Then continue in opposite directions until the ends of the wall are reached, spacing each board as predetermined mathematically. Boards are cut around each opening as required.

6. Fit and nail on the battens, each of which must be carefully plumbed by either using the plumb or carefully measuring from the preceding batten and occasionally checking for plumbness. Be sure to nail only one edge of the batten until the entire exterior finish is completed (to provide as long a shrinkage period as possible).

## CORNER-BOARD CONSTRUCTION AND APPLICATION

The foregoing explanation of siding application did not include information on corner boards, which are often used on external or outside corners in lieu of mitering the siding. In milder climates, where it is not so essential that the exterior walls be made completely airtight (or as nearly airtight as is possible by the use of insulation paper with the siding butted to frame casings and corner boards), corner boards are nailed over the siding as shown in Figure 7.

Corner boards are made in two ways:

1. The cheaper method is to use finished lumber, preferably redwood, western cedar, or some other weather-resisting wood, allowing two pieces of lumber per corner. One piece is narrower than the other so that they measure approximately the same after they are nailed to form the right-angle corner.

2. The better method is to order corner boards which have been sawed from a solid piece of lumber, say 3 by 3 inches.

Careful layout and cutting are needed at the top end of the corner board to make it fit to the cornice member to which it butts. Sometimes, on an open cornice, the corner board must be cut to fit around the end of the rafter that forms the gable of the roof. The top end of each board is first fitted so that it conforms to the actual wall condition. The corner is then assembled on the bench and nailed to the building.

The bottom end of the corner board is marked and cut flush with the bottom edge of the first siding board.

It is best to use 8d cement-coated or galvanized nails for corner boards, as they have good holding power and yet are slim enough to prevent the wood from splitting. The nails should be spaced not more than 12 inches o.c. to ensure a tight job that will not warp or open up.

To construct and apply corner boards, proceed as follows:

1. Read the blueprints to see whether the corner boards are applied before or after the siding.

2. Count the number of external corners and order or select from the lumber pile enough corner-board material for each corner. The blueprint will show the size of the material and also whether a sawed-out or built-up corner is required.

3. If a built-up corner board is specified, be sure to construct it so that the lap joint is in the less conspicuous place; in other words, so that the lap joint will not show on the front elevation of the house.

4. Determine the *approximate* length of each corner board; add 6 to 8 inches and cut two pieces to this length. Place each piece into position to determine what shape to cut the top ends to make a proper fit.

5. Lay out and cut the top ends as needed.

6. Determine exact length and cut off the bottom end of the corner board to line up with the siding.

7. Assemble on the bench.

8. Nail corner boards to place.


### PREPARING FOR AND APPLYING SHAKES, GENERAL INFORMATION*

Natural cedar shingles have adorned the roofs of American homes since pioneer days; the use of cedar shakes for a wall covering is becoming increasingly popular. A combination of siding and shakes, when architecturally correct, provides a beautiful house exterior.

Prestained, processed shakes are manufactured from genuine cedar shingles. The shingles are remanufactured to ensure that the edges of each shingle are parallel and vertical (at right angles) to the butt. The surface is then "vertically combed" to give it an appearance similar to the traditional "hand-split" shake. The shakes are stained in a variety of pleasing modern colors.

Undercoursing shingles are often used to increase the insulation value

---

* The following information is based on "A Handbook for Successful Building with Stained Cedar Shakes," The Stained Shingle and Shake Association, Seattle, Wash. Used by permission with appreciation by the author. (Pages 188–194.)

of the walls and to double the thickness of each "course" of shakes. Since the undercoursing is completely concealed on a finished wall, a very low grade of shingle can economically be used. The stained shakes are then applied over the undercoursing shingles. No space is allowed between adjacent shakes because the edges are parallel and fit tightly together. The vertical lines between shakes blend in with the vertically combed texture (see Figure 8).

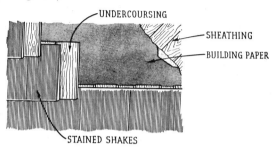

**Figure 8.** Details of shake application using undercoursing.

Window and doorframes must be set before applying the shakes to give the proper finished appearance where the shingles meet the outside edges of the frame casings. Unless the frame has been designed for a shake interior, it will be necessary to make the casings thick enough to "receive" the shakes and form a flush finish.

**Note:** This procedure is always necessary when applying shakes over an existing siding exterior. Obviously, a house planned with a shake exterior will have the exterior face of the frames properly designed to receive the shakes.

Shakes are made in two standard lengths, 16 inches and 18 inches. The amount of shake to be exposed is determined by the height of the wall and the desirability of blending course lines with the top and bottom of the window and doorframe trim.

***Shake-course Rod Layout.*** A rod layout, similar to that described for siding, is a necessity when preparing to apply shakes to the exterior walls of a house. The procedure is as follows:

1. Determine the over-all wall height by selecting the dominant wall of the house (usually the front) and measuring from a point 2 inches below the top of the concrete foundation to the soffit line of the cornice, or any other point where the wall surface and the underside of the roof surface meet (see Figure 9).

2. Divide this total-height measurement into equal parts, as nearly as possible, allowing 12-inch courses for 16-inch shakes and 14-inch courses

for 18-inch shakes. The 12- or 14-inch exposure can be modified to make the shake butts line up with the window and doorframes.

3. Select a suitable piece of rod material, say 1 by 1, and step off the exposure measurement, using a pair of dividers to ensure accurate work.

4. Transfer marks from the story pole to all corners and angles of the house and to all window-frame casings. It is presumed that the insulation paper is in place prior to marking the course locations (see Figure 10).

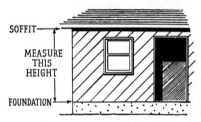

**Figure 9.** Measuring for shake-course rod layout.

**Figure 10.** Transferring rod marks to wall.

## SHEATHING

The sheathing to which the shakes are to be nailed can be either solid or spaced, depending somewhat upon the climatic conditions. Houses of the better class, however, are always sheathed solid, irrespective of geographical location.

Building paper should be used *over* solid sheathing and *under* spaced sheathing to reduce heat transmission through the wall to a minimum.

***Laying Out for and Nailing Spaced Sheathing.*** The first sheathing board is nailed so that its lower edge is flush with the top of the concrete foundation. The second and succeeding sheathing strips are spaced so that the course lines (lower edges of shakes) fall 2 inches below the center of each strip. This method can be reduced to a simple rule: the distance from the bottom of the first sheathing strip to the bottom of the next sheathing strip is 2 inches *less* than the shingle exposure. The distance from the lower edge to the lower edge of all succeeding strips is made the same as the shingle exposure. The top sheathing strip is placed tightly up against the cornice member from which the total wall height was measured (see Figure 11).

Strip sheathing should be nailed with 8d box or 8d common nails; two nails should be used per stud bearing (see Figure 12).

*Laying Shakes on Strip Sheathing.* The undercourse material may be either a low-grade shingle or an insulation backing board, 16 by 48 inches in size. The first or bottom course of shakes requires two undercoursings to make it the same thickness as the succeeding courses. Proceed as follows:

1. Nail on the double row of undercoursing, keeping the lower edges even with the lower edge of the first sheathing strip.

2. Nail on the shakes so that the lower edge is 2 inches below the lower edge of the first sheathing strip. In order to maintain a straight and true bottom edge, strike a chalk line on the undercourse shingles to represent the top edge of the shakes. This can be done because the shakes have been accurately cut to an exact length (see Figure 13).

Nailing must be carefully done. Each shake must be held in place with at least two 5d galvanized

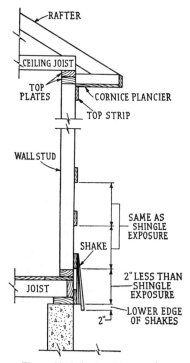

Figure 11. Strip-sheathing spacing.

nails, placed 2 inches above the butt or exposure line and ¾ inch from each edge of every shake. Shakes of greater than 5-inch width require additional nails, spaced no more than 3 inches apart.

Figure 12. Nailing strip sheathing.

Figure 13. Starting first row of shakes.

**Note:** If the wall is solid-sheathed, the same procedure can be followed, except that the first row of double undercoursing must be kept even with the bottom edges of the sheathing. This edge should be sighted to be sure it is reasonably true and straight. The first course of shakes is kept straight by snapping a chalk line to represent the top ends of the shakes, as described in step 2 above.

To lay succeeding rows of shakes, proceed as follows:

1. Select a good straightedge, say a 1 by 4 piece of shiplap, and tack it lightly along a chalked line snapped to represent the lower edge of the undercourse. The rabbet in the shiplap automatically makes a guide for the shake course, which will then project ½ inch below the undercoursing (see Figure 14).

**Figure 14.** Using rabbeted straightedge guide.

2. Nail on the undercourse shingles or insulation board, spacing them approximately ⅛ inch apart and resting them on the upper edge of the straightedge.

3. Nail on the shake course, resting the lower ends on the rabbeted edge of the shiplap and laying them together without spacing.

4. Continue this process until the wall is finished.

*Laying Shakes over Insulating Sheathing.* When insulation sheathing is used, a different method must be followed to provide gripping strength for the nails, as there is no nail-holding power in a piece of insulation sheathing. Figure 15 illustrates the method.

Continuous rows of ⅜- by 1½- by 48-inch wood lath are properly spaced and nailed at each stud. The *lower* edge of the lath is kept ½ inch above the shake-exposure line. To illustrate, if the exposure is 12 inches, then the laths are placed 12½ inches from the lower edge of the shake; if the exposure is 11¾ inches (because the spacing has been equalized to make the top

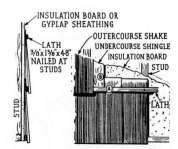

**Figure 15.** Laying shakes on insulation sheathing.

row the same as the other rows), then the lath will be located 12¼ inches up from the bottom edge of the shakes. The *top* edge of the lath serves as a rest for the *lower* edge of the next row of undercoursing shingles. Proceed as follows (no building paper is required):

1. Nail a continuous double row of lath (two laths) to the mudsill, keeping the row true and straight. (Two laths are required for the "starter" row to make the bottom course as thick as succeeding courses.)

2. Nail on a double row of the undercoursing shingles, resting them on the top edge of the lath strip.

3. Determine location of the top end of the shake course, allowing the shakes to project 2 inches below the top of the foundation wall. Snap a chalk line at the point.

4. Lay the shake course, keeping the top ends even with the chalk line.

5. Snap a line on top of the shake course to locate the top edge of the next lath strip and nail on a continuous row of lath. Be sure to allow for the shakes, which project ½ inch below the lath strips.

6. Continue as before: lay the undercoursing shingles; then snap the chalk line and lay the shake course.

7. Repeat this process until the job is completed.

**Outside Corners.** Outside corners are alternately lapped. (This procedure is commonly called "lacing.") Allow one outercourse to protrude slightly past the corner, butting the shake on the other side of it. Trim the protruded corner with a knife or plane. This edge will then need to be touched up with a matching stain.

Another method used for exterior corners is to apply a sawed-out corner board before applying either undercoursing of shakes. The shakes are then fitted against the edge of the corner board (see Figure 16).

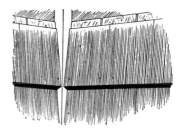

Figure 16. Corner-board construction when shakes are used.

Figure 17. Inside-corner construction when shakes are used.

**Inside Corners.** Inside corners are fitted in the same manner as siding. A suitable strip, approximately 1½ inch square, is nailed into the angle to receive the undercourse and shakes with a minimum of fitting (see Figure 17). The last piece of shake required to fill in the end of each row must be carefully cut to fit the space exactly. The strip must be stained to match the prestained shakes. This should be done after it is nailed into the corner but before the undercoursing and shakes are applied.

*Summary.* Figure 18 provides a quick over-all summary of the various operations required to cover an exterior wall of a house correctly with vertically combed cedar shakes.

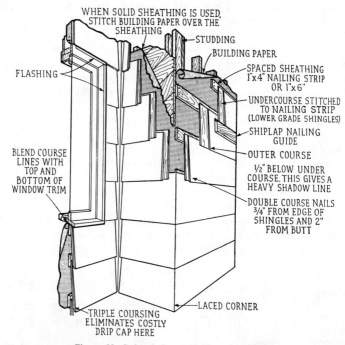

**Figure 18.** Cedar-shake application at a glance.

**Note:** The preceding explanation has described in detail the operations to follow when preparing the exterior walls of a house for shakes and applying the undercoursing shingles and the outercourse of shakes. In mild climates, however, undercoursing is often omitted. The method to follow if shakes only are used is basically the same. Shakes can be nailed directly to either solid or stripped sheathing.

The stripping-lath method must always be followed when insulation sheathing is used because of the non-nail-holding power of this kind of insulation board.

## WATER TABLE

On many houses a combination of siding and shakes is used on the exterior walls. This results in exceptionally pleasing designs, for the siding gives a "horizontal-line" effect while the shakes provide a "vertical-line" effect.

At the point on the wall where the siding leaves off and the shakes begin (often the window-sill line), it is necessary to provide a specially milled detail item known as a water table. A piece of bed molding is fitted under the water table to close up the siding joint (see Figure 19).

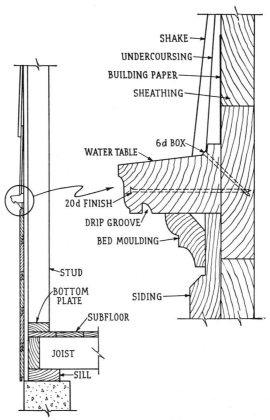

**Figure 19.** Water table.

Water table is made from weather-resisting lumber, such as redwood, with the top face slanted down a little so that rain water will not remain on the surface (hence its name).

The height of the water table is carefully determined from the siding rod, which must be laid out to include the shake spacing. If the window-sill line is used as the line of the water table, then the siding rod must be laid out *downwards* from that point, and the first siding board located accordingly.

Water table must be nailed on perfectly straight and level so that it will be parallel to the siding and will make a correct start for the shake layout. Water table is often at approximately eye level; this is another reason why it must be very carefully applied.

Miter joints, which are used for all exterior corners and inside angles, should also be used when it is necessary to join two pieces of water table together on a long wall, for a miter joint will not open up to the degree that a butt joint will. The miter joint should be made over a stud (see Figure 20).

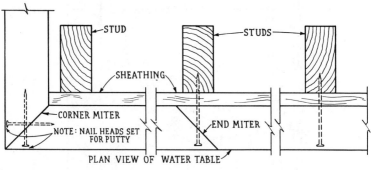

**Figure 20.** Water-table miter joints.

To lay out for and apply water table, proceed as follows:

1. Check the blueprints for the location of the water table.

2. Lay out a combination siding and shake rod.

3. Locate on the wall the exact position of the top edge of the water table.

4. Snap a chalk line on all elevations where the water table is to be used. (Shakes may be only specified on certain elevations.)

5. Select water-table lengths for each wall space to eliminate as many joints as possible.

6. Use a miter box (a metal box is preferred, but a wood miter box made to suit the water table will be satisfactory) and miter one end for an outside corner.

**Note:** The difficulty with a wood miter box is that the saw cuts gradually become worn and out of true. When this begins to happen, simply cut a fresh miter cut.

7. Holding the piece to place, mark where to cut the other end so that it will fit against a stud.

8. Plane this end smooth with a block plane and nail to place, holding it to the chalk line. Ordinarily a 20d finish nail is required for water table because of its width (usually 2¾ inches). One nail is required per stud. Sometimes, in lieu of 20d nails, 6d box nails are toenailed through the top inside corner. These nails will later be covered with the shakes; hence a flat-headed nail can be used to secure additional holding power.

9. Continue until all water table is in place. Care must be used not to bruise the exposed edges of the water table or to damage the outside miters.

**Note:** For the purposes of weather protection, water table should be painted with a coat of priming paint either before fitting it to the wall or as soon as it is in place.

10. The bed mold under the water table is fitted and nailed in place after the siding job is completed.

## CORNICE WORK, GENERAL INFORMATION

A cornice is the decorative unit required to form an attractive finish at the junction of the top of a wall and the lower end of a roof and in a roof gable (see Figures 21 and 22).

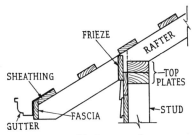

**Figure 21.** Open cornice.

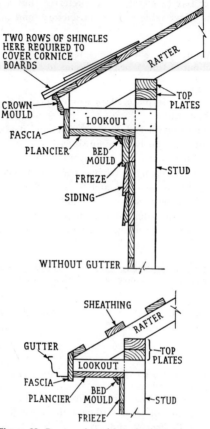

**Figure 22.** Box cornice with and without gutter.

*Cornice Design and Types.* For residential work a cornice may be merely a piece of molding placed to provide a finish edge to the roof, or it may be composed of a number of pieces of finish material. The design of the cornice, when properly executed by a competent architect, adds much to the exterior appearance of a house. Metal gutters are often incorporated into the design.

Cornices can be classified into two divisions: (1) open, in which rafter ends and roof sheathing are exposed, and (2) the box cornice, in which the roof construction is "boxed in" (hence the name) and thus entirely concealed (see Figure 22).

Standard moldings include the bed mold and the crown mold (see

Figure 23). These moldings are milled in different widths and designs—a 2-inch bed mold, a 4-inch crown mold, etc.

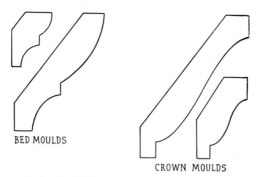

BED MOULDS

CROWN MOULDS

**Figure 23.** Standard bed and crown molds used in cornice work.

**Note:** The term *sprung mold* is often applied to these moldings since they are made from 1-inch lumber in such a way that they have the appearance of being sprung away from the wall and look like a solid molding.

*Building-ordinance Requirements.* Building ordinances covering residential buildings have little to say about the cornice since, structurally, it is not a part of the building. The construction of a cornice, however, should conform to the principles of good framing and result in a piece of work which will not sag.

**Note:** In some localities the building ordinance has definite regulations covering cornice work for such public buildings as stores and schools. In earthquake localities a projecting cornice is not permitted.

*Craftsmanship.* A completed cornice tells much about the craftsmanship of the carpenter who did the work. Anyone who glances quickly along the edge of the roof should see a cornice edge that is straight in both a horizontal and vertical direction. The skill used in making the top plate of a framed wall true and straight simplifies the work of making a straight cornice edge. Care used in selecting straight stock for rafters also shows up in the cornice edge, for a rafter with an excessive crown results in a rafter end that drops below the other rafters, making a low spot in the cornice edge.

*Joints.* Miter joints are required to join various cornice members at exterior corners and at inside angles. Butt joints are required when constructing a cornice longer than the maximum length of a piece of finish cornice lumber (usually 20 feet). These joints must be accurately made.

*Lumber.* The lumber used for cornice members is of the same kind as used in siding work. Redwood, western cedar, and southern white cedar are selected because of their weather-resisting qualities. The underside of a box cornice, however, is subject to little, if any, sun exposure. Redwood and cedar are used here because of their fine working qualities.

### OPEN-CORNICE CONSTRUCTION

The finish material required for the overhang part of an open cornice includes (1) the frieze—material cut in between the rafters to close up the space between the top plate and the underside of the sheathing, (2) the fascia—finish material nailed on at the ends of the rafters—and (3) the exposed sheathing (see Figure 21).

*Overhang.* To provide a suitable paint surface, the portion of the rafter extending below the plate line, called the overhang, should be planed smooth prior to raising the roof. On houses that have a stained exterior the rafter ends and the exposed sheathing are left rough. On some houses with wood shingles the cornice-roof sheathing is so spaced that a part of the underside of the shingles is exposed.

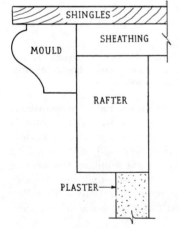

Figure 24. Mold cornice-rake finish.

After the first board is on, the remainder of the cornice sheathing should be left off the cornice area until the cornice edge is straightened. This procedure permits the carpenter to have "stand-up" room while fitting and nailing in the frieze board.

*Rake or Gable-end Cornice.* A cornice constructed to form the finish of a gable end of a roof is called a "rake." Sometimes it is constructed by using a single piece of molding, as shown in Figure 24, or by nailing another rafter on top of the exterior siding, as shown in Figure 25. Since this rafter, called a *verge rafter* or *barge board,* is entirely exposed, it should be S4S and carefully selected for straightness and quality.

The verge rafter needs only a top, or ridge, cut and a bottom, or tail, cut. A strip of 1 by 2 S4S stock nailed on top of the verge rafter and projecting about 1 inch beyond its face covers up the ends of the sheathing boards and forms a suitable finish. The shingles are projected ½ inch beyond the edge of this strip. The rafter is nailed with 16d finish nails through the siding into the studs. Nails should be staggered to hold both edges of the rafter tight to the wall.

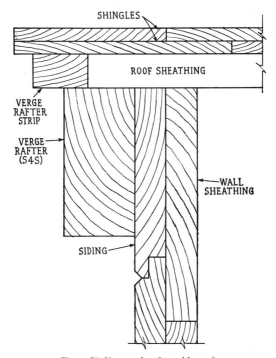

**Figure 25.** Verge rafter for gable roof.

*Straightening a Cornice Edge.* To straighten a cornice edge, proceed as follows:

1. Using a chalk line, test the rafter ends for alignment. "Long" rafter ends must be cut off.

Note: Sometimes a carpenter will let the rafter ends run "wild" until after the roof is raised; then a line is drawn and snapped at the correct location, and the rafters are cut on the chalk-line mark. This ensures a perfectly straight horizontal cornice edge.

2. Bevel the lower edge of the first sheathing board to provide a plumb joint when the board is nailed on top of the rafters. The correct angle can be quickly established by means of a T bevel.

3. Nail on the *first* sheathing board, keeping the lower edge perfectly straight by making it flush with the rafter ends.

4. Sight the sheathing board for "high" or "low" spots occasioned by excessive rafter crown or imperfect rafter-seat cutting.

5. Raise the low rafter by means of a pinch bar and drive a wedge under the rafter-seat cut until the end of the rafter is in true alignment. The "high" rafters must be raised with the bar, the top plate chiseled under the seat cut, and then the rafter renailed to its place.

*Frieze Board.* Cut in the frieze board between the rafters, either using it as separate pieces or, if the board is wide, notching it to suit each rafter and then carefully pushing it up into place and nailing with 6d or 8d box nails. The top edge of the frieze board can remain square, for its lower edge will fit tightly to the sheathing board. Toenail this edge to the rafters.

*Fascia Board.* Fit and nail on the fascia board to the ends of the rafters, nailing into the sheathing edge. Butt joints are used for end joinings. Since the fascia is exposed to much sunlight, 6d and 8d box nails are used to make sure it will stay in place.

*Sheathing.* Nail on the remainder of the cornice sheathing, making it either solid or spaced as required.

*Gutter.* If a gutter is to be used, the sheet-metal man should do his work before the scaffold is removed (see Figure 21).

## BOX-CORNICE CONSTRUCTION

A box cornice requires lookouts nailed at right angles to the studs and projecting beyond the building as much as is necessary to "receive" the plancier board (see Figure 22). The stock used for plancier boards ranges from 1 by 6 to 1 by 18 or 1 by 20. The wide planciers are made from two pieces of finish lumber; the space between is covered with a small molding. Modern design favors the narrower planciers.

*Alignment of Rafters.* The lower ends of the rafters (tail cuts) must be checked for vertical and horizontal alignment. This is done as explained for the open cornice (steps 1 to 5, page 201), except that, if no gutter is used, the first sheathing board projects beyond the plumb cut of the rafter an exact amount to form a nailing surface for the back edge of the crown mold. The projection measurement can be determined only by means of a full-sized layout drawing, as illustrated in Figure 26.

**Note:** The use of a fascia-board crown mold and projecting sheathing, as shown in Figure 26, is also applicable to the open type of cornice.

*Lookouts.* (See Figures 22 and 26.) Lookouts are made from 1 by 4 or 2 by 3 stock. A guideline should be snapped on the face of the wall studs to make the underside of the lookouts perfectly straight. The lookouts must also be nailed level to the sides of the rafters and studs. To ensure a true cornice line, the lookouts should be cut in a cutting box to *exactly* the same lengths.

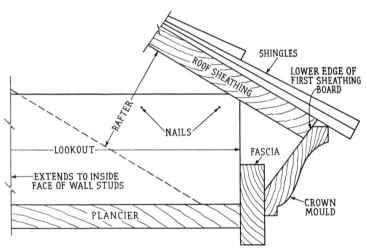

**Figure 26.** Relation of sheathing to crown mold on cornice edge.

The height of the lookouts is determined by the pitch of the roof and the width of the plancier. A full-size cross-section layout should be made in order to determine the exact position of the lookouts. (A layout board can quickly be made from scrap subflooring material.) The layout will also show where to cut the ends of the rafters and how much the sheathing should project.

Lookouts are cut long enough to provide a little more space between the face of the wall and the tail cut of the rafter than the measured width of the plancier material. This will simplify fitting the plancier, as the joint next to the building is covered with a molding. The lookouts are nailed to the exact position determined by the tail cuts at the ends of the rafters.

**Fascia Board.** The fascia is nailed to the ends of the rafters and serves as a guide to keep the outside edge of the plancier straight.

The location of the top edge of the fascia is determined by the cornice design. If a crown molding is used, then the fascia board need only be high enough to "receive" the crown molding. Be sure, however, that a good nailing surface is provided (see Figure 26). After the plancier is in place, the fascia is nailed to it with 8d box nails, spaced at least 12 inches o.c. The lower edge of the fascia projects below the face of the plancier about ½ inch to eliminate a flush joint. Hence the exact width of the fascia must be determined prior to fitting and nailing it to place.

If a gutter is used, the top edge of the fascia board is beveled to suit the pitch of the roof and nailed against the lower edge of the first sheathing board, which has also been correctly beveled to conform to the slope of the roof. This will ensure a tight-fitting joint (see Figure 21).

**Plancier.** The outside edge of all plancier boards must be carefully checked for straightness and, if necessary, jointed until straight. This is important, for the right-angle joint formed by the fascia board and the plancier is always visible. The plancier boards should also be checked to be sure that there are no short "bumps" in the flat of the board. Such boards should not be used.

The inside edge of the board needs no attention since the right angle formed by the plancier and the frieze is covered by a molding.

All end joints must be centered on a lookout. To eliminate excessive joints, avoid using short pieces of plancier. Block-plane all end cuts so that the joints will be smooth.

Erecting a plancier is a two-man job, with one carpenter holding up one end of a board while the other man does the nailing.

If more than one board is required to make the plancier the required width, the same procedure is followed, but no attention need be given to the parallel joints between the boards if the blueprint calls for an astragal molding to cover this joint.

**Frieze Board.** The frieze board on the face of the walls of a house varies in size and design. On a wood-siding house the right-angle joint between the plancier and the siding is covered with a small bed molding. On a plastered exterior the box cornice is completed before the plastering is done; hence the plancier is fitted against the face of the studs, the plastering is done up to the plancier, and a satisfactory joint results.

*Cornice Returns.* When a box cornice is used on a gable roof, it is necessary to construct a cornice "return" on the gable wall, as shown in Figure 27. In effect, a set of tail rafters is nailed to the gable wall to permit the cornice to be "carried" around the corner from 12 to 24 inches. This little piece of shed roof is shingled the same as the regular overhang area, and the plancier forms the finish on the underside.

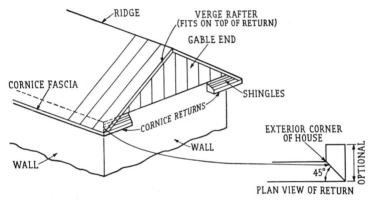

Figure 27. Box-cornice "returns."

*Moldings.* Moldings, as illustrated in Figure 23, are fitted and nailed on the face of the fascia board and into the angle formed by the wall and plancier. The crown mold is used on the fascia and the bed mold on the wall. It is best to use 6d box nails, locating them at least every 12 inches. The nails for the bed mold are driven at right angles to the back bevel of the molding, thus ensuring a tight fit on both edges of the molding. The crown molding is nailed at right angles to the fascia and also at right angles to the lower edge of the first sheathing board.

*Joints.* The miter joint is used for exterior and inside angles; the butt joint is used when two plancier boards are butted together. The miter joint should also be used when two pieces of molding are to be butted together, as is done on water table (see Figure 20).

*Nailing.* Box nails are used for all cornice members to give proper holding power, for a finish nailhead will not hold a board that is constantly exposed to the weather, even if it is kept well painted. Both 6d and 8d box nails are suitable sizes. The plancier is nailed about ¾ inch in from each edge; fascia nails should be spaced about 12 inches apart.

***Box Cornice for Gable Ends.*** A simplified type of box cornice for a gable roof is shown in Figure 28. The plancier is supported by the sheathing boards, which extend beyond the face of the wall the width of the plancier board. A fascia board covers the ends of the sheathing. The lower end of the gable box cornice is fitted to and rests on the cornice return, as illustrated in Figure 27.

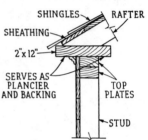

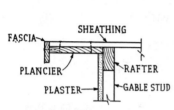

**Figure 28.** Simplified box-cornice rake construction.

**Figure 29.** Framed box cornice.

**Note:** A simplified form of box cornice has been developed in connection with mass-housing projects. A piece of selected S4S framing material is nailed on top of the top plates of the exterior walls, projecting beyond the exterior face of the wall to form the plancier of the cornice. The same piece, if wide enough, can also be projected beyond the inside face of the exterior wall to form plaster backing.

Only exterior walls parallel to the ceiling joists need the plaster backing. The ends of the ceiling joists which rest on the supporting walls must, however, be notched to fit over this "framing-member plancier" (see Figure 29). The framing material is 2 inches thick and varies in width from 8 to 12 inches.

***Summary.*** To construct a box cornice, proceed as follows (a suitable staging is assumed):

1. Test the rafter ends for alignment by means of a chalk line. Long rafter ends must be cut off.

2. Strike a line on the wall to locate the lower edge of the lookouts. The exact location is determined by a cross-section cornice-layout sketch.

3. Nail on lookouts to the line on the wall, keeping the outside ends flush with the rafter tail cuts.

4. Check the blueprint for the shape of the cornice profile: (*a*) with gutter or (*b*) without gutter (Figure 22).

5. If there is no gutter, nail on the first sheathing board, projecting it beyond the rafter tail cut the amount predetermined in the cross-section sketch.

6. If a gutter is used, bevel the edge of the first sheathing board as indicated by the cross-section layout. Then nail the board to place, keeping the sheathing edge flush with the rafter tail cut.

7. Fit and nail on the fascia, locating it vertically so that it will project ½ inch below the plancier board.

8. Cut and nail on the rafter returns in the gables.

9. Fit and nail on the plancier.

10. Fit and nail on the frieze.

11. Fit and nail on the moldings: bed molding in the wall-and-plancier angle, crown molding on the fascia. Omit the crown mold if a gutter is used.

12. Fit and nail on the cornice sheathing.

## GENERAL INFORMATION ON WINDOW AND DOORFRAME UNIT

Window and doorframes form a very important part of the exterior walls of a house and add much to its finished appearance.

Frames are often considered a mill product and consequently are ordered from a cabinet shop specializing in this type of millwork.

It may be thought that a machine-made product is superior to a handmade product, but it is possible, by using such power hand tools as are available or even by using hand tools only, to construct a frame which is as good as one made by machine. Assembling a frame, in any case, is always a hand-tool job. The handmade frame lends itself to construction in the home workshop better than almost any other type of carpentry involved in building a home. The pieces are not too long, so that the frame, except in rare cases, can be completely constructed within the four walls of a shop. The carpenter may, in fact, make the frame on a workbench, using the subfloor of the house as his workshop.

Considerable technical information is necessary before one begins to construct a frame. A careful study of *all* the frame illustrations, Figures 30 to 54 inclusive, is advisable before starting to lay out and make a window frame. No one illustration can include all the possible variations resulting from different methods of construction, types of frames, sizes of framing materials, materials available as stock items at the local lumberyard or mill, climatic conditions, and so on.

The first step is to determine when the frame is to be set. It makes considerable difference, in the construction of a frame, whether it is set *before* the exterior siding is nailed on (this is usually the case in the Eastern part of the United States) or after the siding is nailed on (a common Western practice). Masonry buildings usually require a different type of frame, particularly if the specifications call for double-hung windows, which use a sash cord or chain and sash weights. (This frame is known as a box frame.)

The climatic condition is a deciding factor, for much tighter installation is necessary in colder climates, and architects design window or doorframes accordingly.

***Standard Sash and Door Sizes.*** It should be emphasized that the sizes of all window and doorframes are based on the standard sizes of doors, windows, and sash manufactured by the sash and door mills and factories. Blueprint dimensions are always given on the basis of these standard sizes, unless it is necessary, for some specific reason, to design an opening which does not follow standard measurements. A basic principle of frame layout is, therefore, *"Follow the sizes as given on the plan, or check with the local mills and lumberyards to determine what standard sizes of doors and sash are obtainable."*

Nonstandard doors and sash can always be ordered, but since this procedure is costly, it is advisable to use standard sash and doors whenever possible. The frame layout procedure, however, is the same in either case: follow the measurements of the opening as given on the blueprint.

**Note:** It is advisable not to order the doors and sash until after the frames are made, for some unforeseen framing difficulty may necessitate changing the frame dimensions.

The term *sash* refers to a single unit such as a casement (hinged) sash, a transom sash, or the upper and lower sash in a double-hung (D.H.) window. The double-hung window is divided horizontally into two parts. Two casement sash in the same frame are called a pair and are divided vertically.

***Types of Frames.*** There are several different types of frames, each of which is designated by the way the window opens. Windows which slide vertically, supported by weights or sash balances, are known as double-hung (D.H. for short). A casement-sash frame is one in which the sash may swing either in or out. In a transom-sash frame the sash is usually hinged at the top. In a stationary frame the sash does not open. Doorframes are made to receive exterior doors of a house and are usually constructed to accommodate both a door which swings in and a screen door which swings out. A combination of two or more frames constructed as one unit is designated as mullion, for two frames, and triplet, for three frames (see Figure 30).

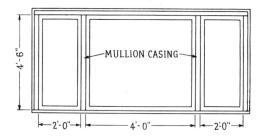

A MULLION DOUBLE HUNG
WINDOW FRAME

**Figure 30.** Mullion- and triplet-frame symbols.

**Note:** Frames constructed to receive the interior doors of a house are often called inside doorjambs, *not* doorframes. Doorjambs are discussed in Chapter 8.

*Plan Symbols.* Figure 31 illustrates the methods used by a draftsman to show a double-hung window, casement window, or door in plan view or elevation. An understanding of these symbols is quite essential to the correct ordering of the materials necessary to construct the frames required for a residential job.*

*Frame Parts.* It is not possible to describe and illustrate fully the many parts that go to make up a specified type of window frame. The problem

---

* See J. Douglas Wilson and Clell M. Rogers, "Simplified Carpentry Estimating," Simmons-Boardman Publishing Corporation, New York, 1956, pp. 75–88, for a detailed description on how to order materials for window and doorframes.

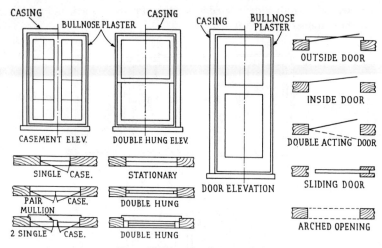

**Figure 31.** Blueprint frame symbols.

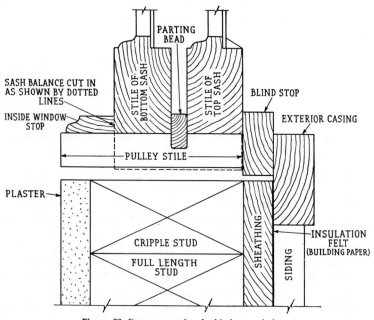

**Figure 32.** Frame parts for double-hung windows.

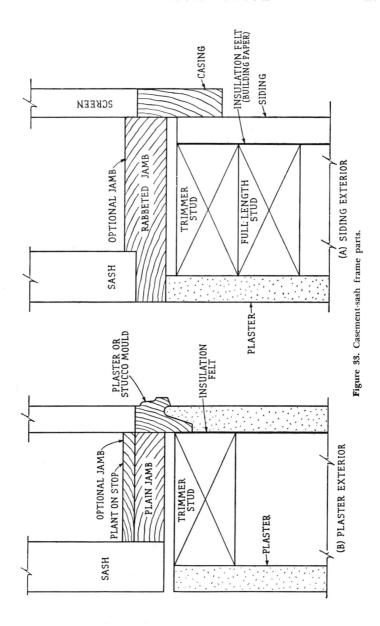

**Figure 33.** Casement-sash frame parts.

becomes even more complex when the various types of frames are considered. The frame parts shown in Figures 32 to 41, inclusive, are the more commonly used stock items that are usually obtainable at any good lumberyard or mill.

The illustrations show several different types of construction, all of which provide for interior plastered walls and require the use of building paper. The various types are as follows:

1. Exterior walls are sheathed, the frames are set, and the siding or shingles are butted to the edges of the casings (see Figure 32).

2. Exterior walls are covered with siding only, and frames are set after the siding is nailed on (see Figure 33A).

3. Exterior walls are plastered (see Figure 33B).

4. Frames are set into a masonry wall (see Figure 34).

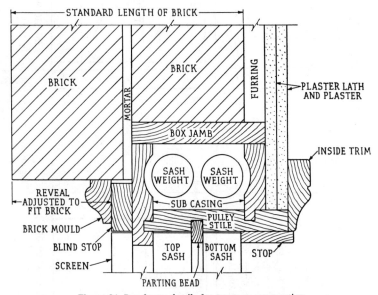

**Figure 34.** Box-frame details for masonry construction.

5. Head casings are provided with a drip cap to receive the rabbeted edge of the siding (see Figure 35).

6. Subsills are used in connection with the water table (see Figure 36).

Most of the different methods of construction are adaptable *to each illustration shown.* These illustrations must therefore be considered as typical only of the various shapes of frame materials that must be

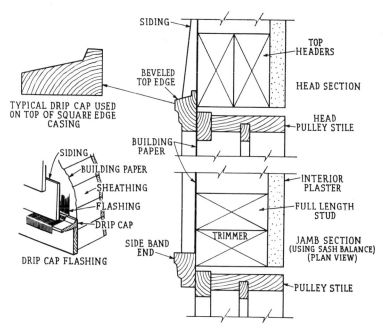

TYPICAL DRIP CAP USED ON TOP OF SQUARE EDGE CASING

SIDING

BEVELED TOP EDGE

BUILDING PAPER

SIDING

BUILDING PAPER

SHEATHING

FLASHING

DRIP CAP

DRIP CAP FLASHING

SIDE BAND END

TOP HEADERS

HEAD SECTION

HEAD PULLEY STILE

INTERIOR PLASTER

FULL LENGTH STUD

TRIMMER

JAMB SECTION (USING SASH BALANCE) (PLAN VIEW)

PULLEY STILE

Figure 35. Head drip cap and side band mold.

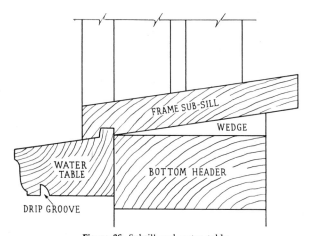

FRAME SUB-SILL

WEDGE

WATER TABLE

BOTTOM HEADER

DRIP GROOVE

Figure 36. Subsill and water table.

SASH WEIGHT     SASH PULLEY     SASH CORD

**Figure 37.** Sash weight, sash pulley, and sash cord.

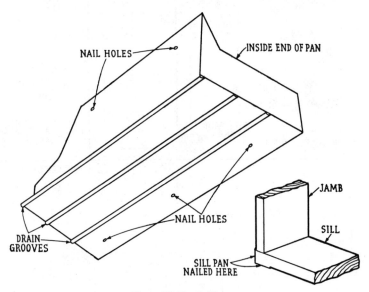

NAIL HOLES

INSIDE END OF PAN

NAIL HOLES

DRAIN GROOVES

JAMB

SILL

SILL PAN NAILED HERE

**Figure 38.** Metal sill pan.

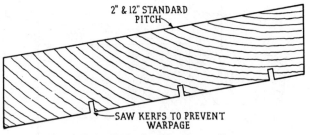

2" & 12" STANDARD PITCH

SAW KERFS TO PREVENT WARPAGE

**Figure 39.** Stock window-frame sill.

selected to construct a frame made to suit a particular wall condition, either framed or masonry.

Each frame consists of two jambs or pulley stiles for the sides, one piece of the same material for the head jamb, and a sill at the bottom. To these basic members must be added blind stops, exterior casings, plaster mold, back band, square stop, apron, pulleys or sash balances, and sill pan, as required. The parts used depend solely on the type of frame being made and the kind of construction around the opening into which the frame is to be set.

Jambs and pulley stiles are made from Douglas fir or white pine. Because of its weather-resisting qualities, redwood is best for exterior casings and window sills. Moldings are also made from redwood.

**Note:** The term *pulley stile,* in a strict sense, is used only in reference to a double-hung window, as it is always grooved to receive the parting bead. The term *jamb* is applied to the casement sash or exterior doorframe. In trade practice, however, the word *jamb* is often used interchangeably with *pulley stile.*

Figure 37 illustrates sash weights and sash pulleys; these are required when constructing a double-hung window frame, unless sash balances are used (see Figure 54).

A metal sill pan is illustrated in Figure 38. This is used to prevent water seepage through the joint. If any moisture does get in, the small grooves will carry it to the outside of the building.

A typical window-frame sill is illustrated in Figure 39. The saw kerfs are cut into the underface of the lumber to prevent warping. Figure 40

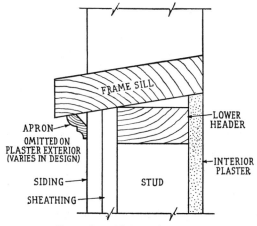

**Figure 40.** Molded exterior apron.

shows a piece of bed molding used as an exterior apron under the window sill.

The parts of a doorframe are shown in Figure 41. Again, it must be noted that the construction shown is not necessarily typical of all framed walls. The combinations of sheathing and siding or plaster are as described above. Sills are often of hardwood, and must be planed flat

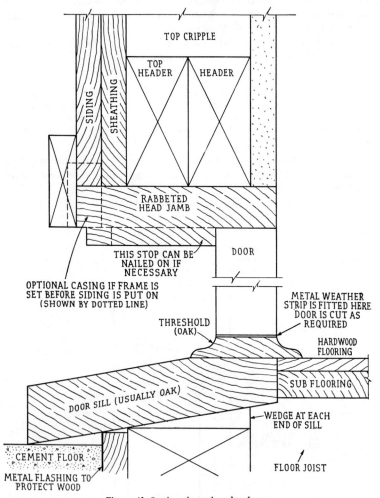

**Figure 41.** Section through a doorframe.

on the part that lies directly under the door to make it possible to use a piece of threshold.

**Note:** On public buildings all exterior doors must open out for purposes of safety in the event of a fire or panic.

There are a number of different methods used to weatherstrip an exterior door or a casement sash that swings in. If weatherstripping is to be used—and it should be used in cold and wet climates—builder's supply houses should be checked for information regarding suitable materials for this purpose which can be obtained locally; the variation in this building item makes it inadvisable to attempt to describe it here.

Jamb stock for either an exterior door or a casement sash (which swings either in or out) varies in thickness from $1\frac{3}{16}$-inch lumber, for a S4S plain jamb, to a piece of 1-inch-net to $1\frac{5}{8}$-inch lumber which has been rabbeted correctly to conform to the exact thickness of the sash or door. The rabbets for the window or door screens must likewise be made to conform to the exact thickness of the lumber from which they are made. Plain jambs require a *plant-on stop,* which automatically forms the screen, sash, and door rabbets. Rabbeted jambs are best because of the increased rigidity of the thicker stock; the separate stop is often used on less costly jobs.

## WINDOW AND DOORFRAME LAYOUT

Three basic measurements must be determined when preparing to lay out a frame: the inside-width measurement of the frame, the inside-length measurement of the frame, and the over-all width of the jamb stock and sill. The last measurement involves consideration of the framed-wall thickness and the kind of finish on both the interior and exterior faces of the wall.

1. *Inside-width Measurement.* The inside-width measurement is determined from the blueprint and is shown in one of two ways: (1) The simplest way (a Western practice) is to include the size, such as 2′0″ or 3′6″ on the floor plan or elevation (see Figure 42A); (2) the Eastern practice is to show the glass size, which may be used to compute the inside-width measurement. If, for example, the glass size is given as 20/18, as in Figure 42B, that means each pane of glass is 20 inches wide and 18 inches high. This 20-inch width measurement must then be increased an amount equal to twice the width of the sash frame from glass rabbet to outside edge. The sash frame is usually 2 inches wide;

hence 4 inches (twice 2 inches) is added to the 20 inches. The inside-width dimension of the frame is thus computed as 24 inches.

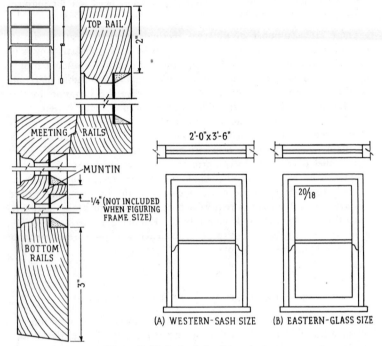

**Figure 42.** Methods of showing sash sizes.

**Note:** If glass sizes are to be used, the exact size of a sash stile should be checked locally before starting to make a window frame, since it may vary in different localities. When muntin bars are used to make a large pane of glass into several small ones, ¼ inch must be allowed for each bar.

If a pair of sash is required for one opening, the width of the frame is made the same as given on the blueprint. The allowance necessary to make the rabbeted joint needed where the two sash meet, as illustrated in Figure 43*A*, is made by the mill sash- and door-layout man.

If two doors (one pair) are required for one opening, then allowance must be made for a T-astragal molding, which is required between the doors to make a satisfactory tight joint (see Figure 43*B*). In this case the thickness of the astragal molding between the doors must be added to the inside width of the doorframe. This is *very important;* otherwise the frame would be too small, and it would be necessary to make

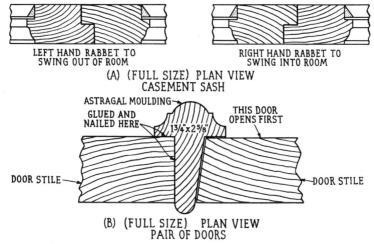

LEFT HAND RABBET TO
SWING OUT OF ROOM

RIGHT HAND RABBET TO
SWING INTO ROOM

(A) (FULL SIZE) PLAN VIEW
CASEMENT SASH

ASTRAGAL MOULDING

GLUED AND
NAILED HERE

1¾"x2⅝"

THIS DOOR
OPENS FIRST

DOOR STILE

DOOR STILE

(B) (FULL SIZE) PLAN VIEW
PAIR OF DOORS

**Figure 43.** Casement-sash rabbets and T astragal for pairs of doors.

the four stiles of the doors narrower, spoiling the design of the door. The astragal is glued and nailed to one door edge, as shown.

Mullion and triplet frames require special width allowances for the mullions (vertical members) which separate the sash, as shown in Figure 44. The thickness and width of a mullion vary even on the same job. Sometimes a piece of 2-inch milled stock, measuring 1⅝ inches between the sash, is used. Again, the mullion bar could be 6 or 8 inches wide, in which case it consists of two separate jambs, the space between being covered by a mullion casing.

The procedure to follow when laying out the width of a mullion (or triplet) frame is as follows: start from one end of the sill or head jamb (being sure to allow for the sill horn) and measure for one sash; then allow for the mullion as detailed and measure for the next sash. Figure 44 shows a continuous head jamb for a triplet frame and an alternate method of using three separate pieces of stock for the head jamb. The separated head jamb allows for a vertical framing member to be nailed between the top and bottom headers in case the opening is wide enough to require additional support for the top header.

Sill horns vary in length, depending on the width of the exterior casing which rests on this part of the sill (see Figure 44). A basic rule to follow when determining the length of the horn is, "Allow as much sill beyond the edge of the casing as there is beyond the face of the casing." To illustrate, if the sill projects 1¼ inches beyond the face

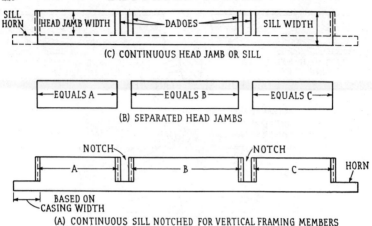

**Figure 44.** Layout for multiple-frame sill and head jamb.

of the casing, it should also project beyond the edge of the casing $1\frac{1}{4}$ inches.

**2.** *Inside-length Measurement.* The inside-length measurement of a frame is a little more difficult to determine, irrespective of which method of stating the sash size is given. The length of the sash must first be determined. In Western practice this may be done by reading the blueprint. In Eastern practice it is necessary to read the length measurement of the glass, double it (because there are two panes of glass), then add 6 inches—2 inches for the top rail, 3 inches for the bottom rail, and 1 inch for the meeting rails. In Figure 42*B,* the glass size is 18 inches. Two times 18 inches equals 36 inches. Thirty-six inches plus 6 inches equals 42 inches or 3′6″. Hence the window shown is 2′0″ by 3′6″. Note that the size shown on the floor plan illustrating Western practice is also 2′0″ by 3′6″. (Muntins are *not* included when figuring frame size.)

The next step after calculating the length of the window or sash is to determine where to measure this length on the frame jamb or pulley stile. It is at this point that a detailed understanding of the type of frame to be made is needed. Figure 45 illustrates the three typical situations that pertain to windows and sash. Figure 46 shows where and how to measure the length of a doorjamb.

DOUBLE-HUNG-WINDOW FRAME. (See Figure 45*A.*) The length of a double-hung-window frame is measured along the inside edge of the parting-bead groove from the inside face of the head jamb to the face

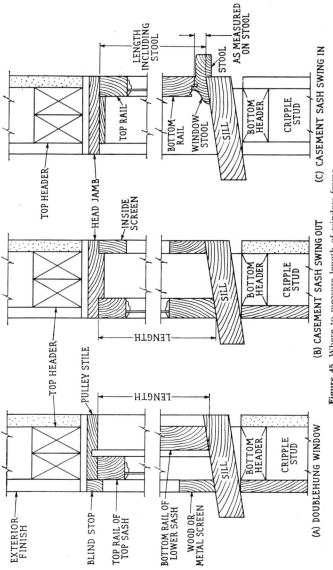

**Figure 45.** Where to measure length of window frame.

of the sill. Proper allowances must be made for the joints at the top and bottom end of the side pulley stile (see page 226 for types of joints).

CASEMENT-SASH FRAME (SWING OUT). (See Figure 45*B*.) The length of this frame is measured at a point on the side jamb which represents the outside face of the sash. This measurement must be taken from the face of the top head jamb that represents the top edge of the sash. On a rabbeted jamb be sure that the measurement is taken from the face of the rabbet in the head jamb.

CASEMENT-SASH FRAME (SWING IN). (See Figure 45*C*.) When determining the length of the swing-in type of casement-sash frame, additional

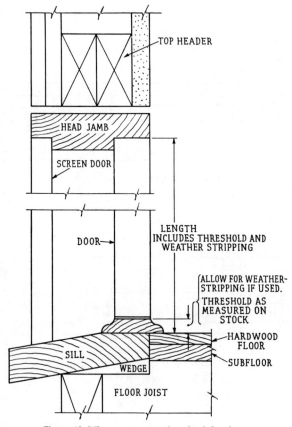

**Figure 46.** Where to measure length of doorframe.

allowance must be made for the piece of window stool which is required under the sash to make a watertight joint (as well as to provide a part of the inside trim). This extra amount of length will vary with the stool pattern; ¾ inch is standard, but the exact measurement must be ascertained before the frame is laid out.

Take the length measurement along the inside edge of the side jamb, from the underside of the head jamb to the face of the sill, and then add the stool thickness. This increased measurement locates the cutting point for the frame sill.

EXTERIOR DOORFRAME. (See Figure 46.) The length of an exterior doorframe is measured along the inside edge of the jamb. Measure from the face of the rabbet in the head jamb a distance equal to the length of the door as given on the blueprint; then add the thickness of the threshold, which is required under the door to give sufficient rug clearance when the door is opened. Thresholds are usually ⅝ inch thick, but the exact measurement should be checked at the local mill or yard before the frame is made.

Note: A standard screen door is always made 1 inch longer than a standard exterior door to allow for the pitch of the sill. This is the standard procedure followed by sash and door mills. Hence only the inside-door length dimension needs to be measured. A screen door for a 6′8″ doorframe, for example, would be made 6′9″ long.

**3. Width of Jamb Stock.** The width of the jamb stock is determined by the kind of wall construction used on the house. Figure 47 illustrates several different possibilities. The actual mathematical process followed to find the width of the jamb stock or pulley stile is really very simple: (1) read the blueprint and specifications to see what kind of construction has been designed; (2) find the *exact* dimensions of each part of the construction: stud width, sheathing thickness, and siding thickness or shingle-wall sheathing and shingle thickness. To this combined measurement add ¾ inch for inside plaster. (Outside plaster is also figured as ¾ inch thick.) The net result is the thickness of the finished wall from interior face to exterior face. This over-all dimension becomes the basic measurement to use when determining the exact width of the jamb stock.

The actual width of the jamb is determined by *subtracting* from this over-all wall measurement the thickness of the various window-frame members, such as the blind stop or plaster mold, and also the thickness of the bullnosed plaster finish if this is used on the exterior walls. The answer obtained equals the width of the jamb.

Note: Plastering, either interior or exterior, is done *after* the frames are set. The inside edges and outside moldings of the frame, therefore, must project

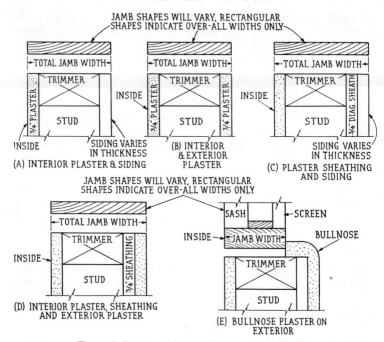

**Figure 47.** Factors which determine over-all jamb width.

beyond the face of the wall the exact amount of the desired plaster thickness. The plasterer then works to the frame, the edges of which serve as a guide to make a true and straight plaster surface of the proper thickness. In some localities the thickness of the plaster is specified by the building ordinance; hence the ordinance must also be consulted before finally settling on the jamb-width dimensions.

Figures 32, 33, and 47 also illustrate the jamb-width factors which must be carefully studied and accounted for when determining the net width of the jamb stock.

Frame stock can be purchased that has been milled with rabbeted edges and grooves to make a more satisfactory watertight joint, as shown in Figure 48. The jamb liner is made in several different widths to permit assembling a frame jamb of the exact width to conform to the wall thickness. This milled material is quite desirable. Carpenters who plan to use it, however, should make a careful check *at the mill* to be sure that materials of the correct width to do a required job are obtainable.

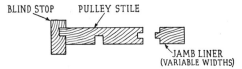

**Figure 48.** Milled joints.

**4. Width of Sill Stock.** (See Figure 40.) The width of the sill stock is based on the wall thickness plus the amount that the sill projects beyond the face of the exterior wall to provide a satisfactory design at the bottom of the frame. It is common practice to buy standard widths of sill stock and rip them to meet the requirements of the framing conditions.

## WINDOW AND DOORFRAME CONSTRUCTION METHODS

The term *construction methods* is used here to describe the various types of joints used at the corners of a frame and the procedures involved in fitting mullions to the head jamb and the frame sill. Such topics as how to install sash pulleys, where and how to cut in a weight pocket, and how to prepare for sash balances will be covered under Special Construction Features. Nailing specifications are included.

*Tools Required.* A good workbench, such as the one illustrated on page 178, is very necessary for frame construction. It is also important to have a pair of sawhorses with a smooth top that will not scratch or mar the finish frame lumber.

Finish tools include finish saws, a backsaw, a set of chisels, a brace and set of bits (for the pulley holes), smoothing planes, a try square, a T bevel, marking gauge, a jointer plane to straighten the jamb stock when necessary, a good 12-ounce finishing hammer—in addition to the standard 16-ounce hammer, a screwdriver of the correct size, and lastly, but most important, a router plane. A miter box is essential, too; if possible, a metal box should be used; otherwise a wood miter box should be carefully constructed for the frame job.

*Frame Joints.* (See Figure 49.) Frame joints can be roughly classified into two groups: (1) the joints used to fit and fasten the jamb members together and (2) the joints used to fit the various frame members which are nailed onto the jamb frame.

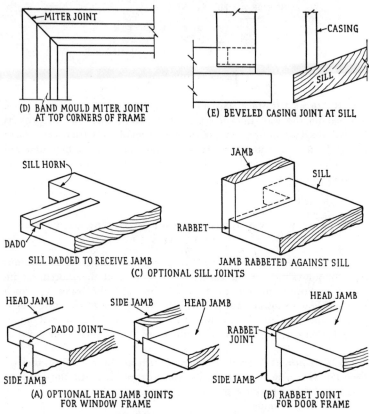

(D) BAND MOULD MITER JOINT
AT TOP CORNERS OF FRAME

(E) BEVELED CASING JOINT AT SILL

SILL DADOED TO RECEIVE JAMB

JAMB RABBETED AGAINST SILL

(C) OPTIONAL SILL JOINTS

(A) OPTIONAL HEAD JAMB JOINTS
FOR WINDOW FRAME

(B) RABBET JOINT
FOR DOOR FRAME

**Figure 49.** Typical frame joints.

The rabbet or dado joint is best for the corner construction of a frame. The width of the dado is made the same as the thickness of the stock which is to fit into the dado. The depth of the dado is usually ¼ or ⅜ inch, unless the jamb stock is made from 2-inch stock, in which case the dado is made ½ inch deep.

The dado joints for the top corners of the frame are made in one of two ways: (1) by cutting the dado into the head jamb (see Figure 44, which shows a continuous head jamb for a triplet window frame); (2) by cutting the dado into the top end of the side jamb (see Figure 49*A*). Either method is correct.

The foregoing explanation also applies to a corner rabbet joint, which is most often used on a piece of rabbeted doorjamb.

*Sill-joint Layout.* The procedure followed in making dado joints for the top corners of the frame is also used for the sill: either cut the dado into the sill, in which case, because of the pitch of the sill, the dado tapers to nothing, as shown in Figure 49*C*; or cut the dado into the lower end of the jamb stock (see Figure 49*C*). It is very important, when dadoing the lower end of a piece of jamb stock or pulley stile to receive the sill, to make the cut to the correct slant. Once the bevel has been established (with the pitch of the sill serving as the angle), a T bevel is used to lay out the correct angle on all the pieces. Sometimes the metal miter box can be locked to a desired position, or a wood miter box may be used to make a cut to the exact slope.

Flat casing joints are sawed on a square-beveled cut to fit the sill (see Figure 49*E*). The top end is cut square to fit the square edge of the head casing (see Figure 50).

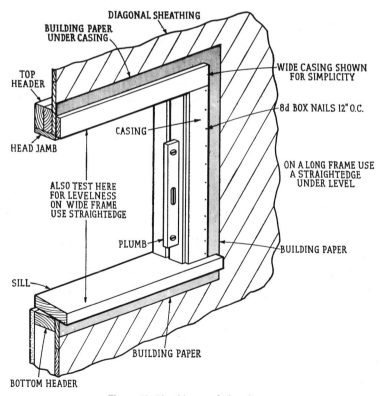

**Figure 50.** Plumbing a window frame.

All moldings are mitered at all corners and bevel-cut to fit the sill. Aprons are return-mitered to make the end of the apron the same profile as its face or they can be cut with a coping saw to make the end cuts the same profile as the face of the apron.

*Cutting Frame Stock to Rough Lengths.* Window and doorframe materials are usually ordered in long lengths which have been selected to eliminate waste. Therefore, when starting to make a frame (or frames), it is necessary to make a cutting list, including two side jambs, one head jamb, and one sill per frame, plus all casings, blind stops, and parting beads. The "rough" length of the frame part is based on the actual inside dimensions of the frame and the amount of additional material required to allow for the joints.

Proceed as follows:

1. For side jambs that are to be dadoed to receive the head jamb and the sill, add 6 inches to the inside length of the frame.

2. For side jambs that are to be dadoed into the head jamb and the sill, allow 2 inches more than the inside length of the frame. For a doorframe include $\frac{5}{8}$ inch for threshold clearance (see Figure 46).

3. For head jambs that are to be dadoed into the side jambs, add 1 inch to the inside width of the frame.

4. For head jambs that are to receive the side jambs, add 4 inches to the inside width of the frame.

5. For sills that are to be dadoed into the side jambs or are to receive the side jambs, add twice the length of the sill horn to the inside width of the frame; then add 2 inches more.

6. For a box-frame head or sill, add to the inside width of the frame twice the amount required for one box (jamb thickness plus pocket width, usually $2\frac{1}{4}$ inches, plus subjamb thickness); then add 2 inches for cutting; for the sill add twice the length of the sill horns (see Figure 44).

7. Rough-cut blind stops and plaster moldings on the same basis as the jambs.

8. Rough-cut side casings to the inside length of the frame plus 2 inches.

9. Rough-cut head casings to the inside width of the frame plus twice the width of the casings plus 2 inches.

10. Parting beads are usually cut to exact lengths after the frame is made. First make a pattern; then, using a miter box, cut the head parting bead and the side parting beads to fit exactly. Rough cutting is not required, since the molding is very small and easy to handle in long lengths.

**Note:** The above rules apply only to a single frame; if a mullion or triplet frame is to be made, then add to all head-jamb or sill measurements the width of each mullion.

*Jamb Layout for Pairs.* The jambs must always be laid out in pairs, as illustrated in Figure 51. If pairs are not made, the result will be too many "lefts" or too many "rights."

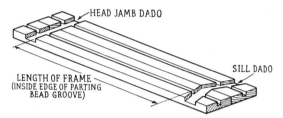

Figure 51. Layout for a pair of jambs.

To be sure that pairs are laid out, proceed as follows:

1. Lay out a piece of jamb stock to the required length.

2. Use this piece as the pattern; place it against another piece of stock, as illustrated, and make the sill cut in the opposite direction.

3. Be sure that each pair is laid out to exactly the same length.

*Checking Head-jamb and Sill Layout.* The length of the head jamb and the sill should be checked, as shown in Figure 52, to be sure they are exactly alike. The same procedure should be followed for all mullion and triplet frames, irrespective of whether the head jamb is continuous or in separate pieces. If all parallel pieces are cut exactly alike, then a square frame will result. If parallel pieces are not the same length, it will be impossible to make a square frame. The results of this type of poor craftsmanship show up when the sash and inside trim are fitted, for much time and effort is then consumed in trying to make perfect miter joints on the trim, the rail and stiles of the sash are not parallel, and the final effect is displeasing to the eye.

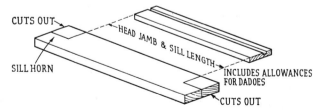

Figure 52. Sill and head-jamb (length) checkup.

***Cutting in a Dado by Hand.*** Dadoes are easily cut in by hand with the aid of a backsaw, router plane, and chisel. Assuming that the layout work has been completed and that the jamb stock is clamped or otherwise held to the bench top, proceed as follows:

1. Tack on a thin strip of wood, say ⅜ inch thick, exactly to one of the lines representing the dado to be cut. Be sure to use small nails and to bend them over to prevent tearing your hands in the work that follows.

2. Carefully saw along the line to the required depth. (The edge of the dado marks should be gauged for depth.)

3. Remove the guide strip and renail in readiness for the next sawing. Be sure that all cutting is done *inside* the dado lines; otherwise the dado will be too wide.

4. Using a wood chisel of suitable width, rough-cut the stock between the two saw cuts. Be sure not to go too deep and not to chisel into the open ends of the dado, lest a split jamb edge result.

5. Set the router-plane bit to the exact dado depth required. Using both hands, carefully push the router along the dado (the action is much like planing), continuing until the bottom of the dado is completely smooth. Be sure that the router is not pushed into the open ends of the dado; in other words, work from both edges toward the middle.

Note: The small nail holes made when setting the guide strips are not objectionable. If care is used, some will come below the sill line or above the head-jamb line, where they will never be seen. The method described above is a very satisfactory alternative to the use of a powered dado saw.

## SPECIAL FEATURES OF WINDOW AND DOORFRAME CONSTRUCTION

***Sash Weights and Sash-weight Pockets.*** Double-hung sash are supported by sash weights fastened to a piece of sash cord which passes through a sash pulley inserted near the top end of the pulley stile and is then fastened to the edges of the sash which have been bored and grooved to receive the cord and knot.

Double-hung windows vary in size and in the kind of glass used—sometimes single-strength (16-ounce) glass and sometimes double-strength (21-ounce) glass; hence each sash must be weighed at the time it is being fitted, and sash weights (see Figure 37) ordered accordingly. This procedure is explained fully in the next chapter.

The space in which the sash weights slide, between the back of the pulley stile and the face of the trimmer stud, must be at least 2¼ inches. A wide interior casing is required to cover this space.

Holes for sash-weight pulleys are made to conform to the pulley to be used. Pulleys vary from a simple pressed-steel type, which is quite durable (see Figure 37), to a very heavy type, more often used on public buildings, which has a brass-finish surface and must be carefully dapped into the face of the pulley stile. The pressed-steel pulley is held in place by means of a 16d finish nail, which is passed through a hole in the center of the wheel axle and on the back side of the pulley stile. This method holds the pulleys tightly in place, one nail only being required for both pulleys. The pulley with the brass face is screwed into the face of the stile.

Pulleys are centered on the sash-run line, as shown in Figure 53, and the top end of the pulley is located about 4 inches below the inside face of the head jamb.

CUTTING A SLOT FOR SASH-WEIGHT PULLEYS. To cut a slot for sash-weight pulleys, proceed as follows:

1. Carefully make a light line to represent the center of the pulley.

2. Measure the over-all length of the part of the pulley which is to be inserted through the pulley stile.

3. Mark crosslines on the center line to represent the ends of the pulley.

4. Select a wood bit which will make a hole just wide enough to receive the pulley. (It will be necessary to measure the pulley to determine the number of the bit.)

5. Carefully bore a series of holes on the center line, being sure that the end holes do not extend beyond the crosslines which represent the ends of the pulley slot. The pulley stile should be clamped to a piece of wood to prevent the wood bit from splitting the back side of the jamb. (The only alternative is to bore from the face of the jamb until the spur of the bit is felt on the opposite face; then turn the piece over and bore from the other side.)

6. Chisel the edges of the slot until they are straight. The slot is now ready to receive a pressed-steel pulley.

7. If a cast-steel pulley is used with a finished faceplate, place the pulley into the hole, mark carefully around the edges of the brass plate, and remove the pulley; then remove the wood within the mark to a depth equal to the thickness of the plate. This completes the slot and dap for this type of pulley.

HOW TO CUT IN A SASH-WEIGHT POCKET. When sash weights are used— and this practice is still considered excellent construction—a sash-weight pocket must be cut into each side pulley stile, as illustrated in Figure 53, to permit the repair of a sash cord which may later become broken through constant use.

The width of the pocket is from the inside edge of the pulley stile

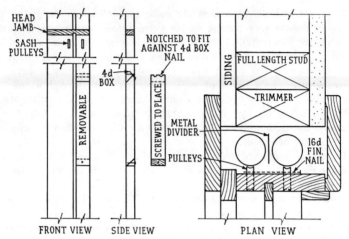

**Figure 53.** Removable weight pockets.

to the parting-bead groove. The length of the pocket depends on the size of the frame, for a larger frame requires a heavier (therefore longer) weight and hence a longer pocket. The average pocket is about 12 inches in length. The bottom end is located about 6 inches from the sill.

The ends of the removable piece of pulley stile which forms the entrance to the weight pocket are cut to a 45-degree angle, as illustrated in Figure 53. Proceed as follows:

1. Lay out the end cuts of the pocket hole, approximately 6 inches from the sill line and at least 12 inches long. Square two light lines across the face of the stile.

2. Using a miter square, make two angle cuts on the edge of the stile as illustrated.

3. Using a backsaw, carefully cut on these lines, sawing from the face of the stile. Be sure not to cut beyond the parting-bead groove.

4. Using a chisel, cut the bottom of the groove to cause the piece to split in the direction of the grain. A finish cut is not necessary since it is always covered by the parting bead.

5. Remove any splinters from this edge.

6. Notch the top end of the removable piece, as shown, to make a square end to provide pressure against the two nails which hold the top end in place.

7. Hold the piece in position so that the lower cut is tight and drive in the two nails, as shown. Because the two saw cuts have shortened the piece, the lower end will now project beyond the face of the pulley stile. This surplus wood must be planed off.

8. Select a flat-headed screw of a suitable size and fasten the lower end to the stile. This screw is removed whenever it becomes necessary to open up the pocket to repair the cord.

*Sash Balances.* With the advent of very narrow inside casings around the inside edges of the window frames, which left no space for the weights to slide in, it became necessary to devise some other means of holding up the sliding sash in a double-hung window. The spring sash balance was the solution to the problem.

Note: It should be pointed out that sash weights can still be used with a narrow inside trim if the wall is of sufficient thickness to permit the use of a box frame. This is the procedure followed on brick buildings. Double-hung windows in many public buildings are installed with sash weights and pulleys. Because of the larger size of the windows, chain is used instead of cord, as it has greater durability and strength.

Sash balances, similar to sash weights, are purchased on the basis of the weight of the sash they are to hold; the metal case, which encloses two springs (see Figure 54), one for each sash, is made the same size, irrespective of sash weight, since the difference is made in the adjustment and power of the springs. All sash-balance holes, therefore, are the same size. For a single installation it is customary to place the balance in the left-hand pulley stile (when viewed from the inside of the house). A double installation (two balances per frame) requires cutting into both pulley stiles.

The cable is made of a very strong and flexible wire with a metal holder on its end, which is screwed to the lower corner of each sash.

CUTTING IN A SASH BALANCE. The pulley stile must be cut as illustrated in Figure 54. Cut a rectangular hole the exact size of the body of the sash balance. To illustrate, for the Acme Twin sash balance (illustrated) the hole is made $3\frac{1}{4}$ inches wide and $6\frac{3}{8}$ inches high. The lower end of the cut is made exactly 2 inches above the center of the frame, as measured vertically. The inside edge of the hole is located by measuring from the outside edge of the pulley stile.

To assure a smooth-running sash, a metal glide is placed on the edge of the sash opposite the balance (see illustration).

Note: The Acme Twin * is the most efficient sash balance ever designed. It requires no adjustment, it is easily installed and weatherstripped, and one balance handles both the upper and lower sash of windows that are not over 16 pounds. The faceplate fits flush in the window frame, the spring mechanism being enclosed in a rustproof case only $\frac{3}{4}$ inch thick. The cutout in which the balance is inserted should be 2 inches above the center line of the frame.

* Descriptive material and illustrations courtesy of the Acme Sash Co., Division of Duplex, Inc., Los Angeles, Calif.

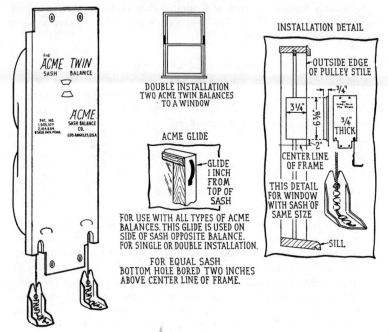

Figure 54. A sash balance.

One Acme Twin handles both sash of lighter windows, while larger and heavier windows require double installation. The Acme Twin is available in sizes to counterbalance sash weighing from 4½ to 32 pounds.

## ASSEMBLING WINDOW AND DOORFRAMES

*Nailing.* It is assumed, in the following explanation, that all jambs and sills have been rabbeted or dadoed ready and that all holes for pulleys or sash balances have been made.

The size of nail to use when assembling a window or doorframe depends on the thickness of the lumber. For all rabbeted and dado joints flat-headed nails are used, since the nails are not seen after the frame is set. For 1-inch jamb stock drive two 8d box nails ½ inch from each edge through the jamb into the head jamb and two 6d box nails 1 inch from each edge through the back side of the head jamb into the side jamb. Wide jambs require 3 nails in each direction. Nailing the joint from two different directions will keep the joint from opening up.

The same directions can be followed for sill nailing, except that the nails driven through the sill into the jamb should be 8d box.

Surface or exposed nailing for flat casings, since they are exposed to the weather, requires flat-headed nails. It is best to use 8d box nails, spacing them not more than 9 inches o.c. If the nailheads are to be covered with putty, then each one must be set about $\frac{1}{16}$ inch below the surface of the wood.

Moldings usually require either 6d or 8d finish nails, depending on the size of the molding. Finish nails are usually used, even though the moldings are exposed to the weather, because a flat-headed nail will tear the molding contour.

Blind stops are surface-nailed with 6d box nails.

Miter joints should be nailed in both directions to prevent the joints from opening up.

Solid mullion bars require 16d box nails at the sill joint.

Parting bead for double-hung windows is fitted into the parting groove but *not* nailed, for it must be removed when fitting the sash. The head parting bead is cut in first; the sidepieces, being fitted under the headpiece, serve to hold the toppiece in place. The groove and bead are milled to give a tight fit. It is customary, however, to use a 4d finish nail and drive one along the outside face of each sidepiece of parting bead and bend it over slightly to prevent the bead from falling out.

In summary, all frames should be well nailed at all corner and face joints (even though they carry no load and no racking strain can be placed upon them), as they are constantly exposed to changing weather conditions.

*Assembling Procedure.* The head, side jambs, and sill of a frame are nailed together first. The frame should then be squared and temporarily braced. To square a frame use a small strip of wood and check the diagonals. For accurate squaring, the rod should be pointed at one end to permit the point to be placed exactly in the inside corner of the frame. The opposing angle is marked with a pencil; the frame is then checked in the opposite direction, and the rod is again marked; the frame is racked slightly, if needed, until both diagonals register exactly the same.

A piece of 1 by 1 or 1 by 2 long enough to reach from corner to corner is nailed on the *inside* edges of the frame to hold the frame until after it is set in the wall.

**Note:** The steel square should be used only on small frames or frames which are made from 2-inch stock. Even then the side jambs must be sighted for straightness as a frame cannot be squared by means of a steel square if any of the frame pieces are bowed; hence the "diagonal-checking method" is the only sure way of squaring a frame.

After the frame is squared, proceed with the cutting, fitting, and nailing of all exterior casings and moldings. All margins should be kept exactly parallel. To locate the blind stop ½ inch beyond the face of the pulley stile, make a gauge block to the exact dimension; this block serves as a nailing guide to assure an exact and parallel projection.

A marking gauge is often used to locate the exact position of a molding which is to be planted on the face of the casing. The gauge line is covered by the molding and hence is not seen after the frame is assembled.

## SUMMARY OF WINDOW AND DOORFRAME CONSTRUCTION

The preceding information covers in detail various phases of frame making. The explanation may make frame construction seem difficult. This is not the case, however. Provided that one selects the methods described, the actual steps required to build a frame are simple, although they do demand a high degree of tool skill.

Assuming that only one frame is to be built, the basic steps are as follows:

1. Determine the type of frame to be constructed: casement, double-hung, or door and whether for frame or masonry walls.

2. Determine whether the frame is single or a mullion or triplet combination.

3. Find out, mathematically, the exact thickness of the wall into which the frame is to be set.

4. Read the blueprint and find out the construction details to be observed: set before siding, set after siding, plastered exterior, bullnosed plaster returns, etc.

5. Read the blueprint and find out or compute the inside dimensions (sash size or door size) of the frame.

6. Select the jamb stock, reading the specifications for thickness dimensions.

7. Cut two side jambs to a rough length, approximately 6 inches longer than the sash or door length.

8. Cut the head jamb to rough length, 1 inch longer if the head jamb is dadoed into the side jamb, 2 inches longer if side jambs are dadoed into the head jamb.

9. Cut the sill piece to rough length, adding sufficient stock to make the sill horns.

10. Place the workbench in a good, solid position, sweep clean, and set all protruding nails so that the surface is clean and smooth.

11. Select a good pair of sawhorses and, if necessary, nail on a smooth piece of 1-inch stock to provide a smooth working surface, free from roughness.

12. Lay out the side jambs to exact length, being sure to make a pair of jambs and not two "lefts" or two "rights."

13. Lay off the head jamb to exact length.

14. Lay off the sill to exact length, using the head jamb to check by.

15. Cut and router out all dadoes and rabbets.

16. Assemble the jamb frame.

17. Square and brace the frame.

18. Cut and fit all other parts as required.

**Note:** To provide additional weather resistance, frames should be back-primed before setting. This process is simply one of giving a good coat of priming paint to the parts of a frame that are not exposed, that is, the back sides of the jambs, casings, sills, etc.

### SETTING WINDOW AND DOORFRAMES

*Setting a Window Frame.* The procedures followed in setting a frame are the same whether the job is done before or after the siding is on or when the exterior wall is plastered. The procedure is slightly different when a subsill frame is used in connection with the water table because this member automatically establishes the height of the window frames. Frames set in brick walls involve an entirely different process, for the frame setting must be coordinated with the bricklayer's work.

The story pole is quite important in frame setting since it shows the exact height of the head jambs of the window frames, which is usually the same as the doorframe height. If building paper is used it must be nailed on prior to setting a frame. The paper is cut into 10- or 12-inch strips and nailed around all edges of the opening.

To set a window frame, proceed as follows (two men are required to do the job):

1. Measure on the story pole the height from the subfloor to the underside of the head jamb (usually inside-door height plus $\frac{1}{2}$ inch for rug clearance plus thickness of the hardwood floor). Cut two rods, say 1 by 2 to this length. These pieces are used to hold the frame to its height position while setting it in the opening.

2. Set the frame into the hole, working from the inside of the building (this avoids the need for a scaffold) and rest it on the rods, as shown in Figure 55.

3. Center the frame in the opening by measuring the spaces at each side of the frame and equalizing them.

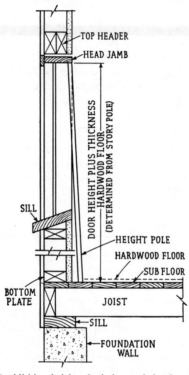

**Figure 55.** Establishing height of window and doorframe for setting.

4. Hold the frame tightly against the wall surface; then nail through one top corner of the outside casing, using an 8d box nail. Do not drive the nail "home" until further checking is done.

5. Repeat this step on the opposite top corner, but check the head jamb for levelness before driving the nail. It must not be assumed that the floor area on which the two rods that hold the frame up are resting is exactly level; hence all frame heads must be carefully checked for levelness. When the head is level, then drive the nail part way in.

6. Next check one side of the frame to be sure it is plumb and drive a nail in the lower outside corner of the casing just above the frame sill (see Figure 50). Do not drive it home.

7. Check the other side jamb for plumbness and tack the casing to the wall.

8. Remove the rods and check the sill for levelness.

**Note:** The reason for tack-nailing on four corners of the frame and checking for levelness and plumbness on all four sides of the frame is to make sure that the frame is set exactly square. It must not be assumed that the frame is perfectly square; if it is found that the frame will not plumb on both of its sides, adjust it slightly by removing the lower nails and even up the discrepancy.

9. A second way of checking is to test the frame with the framing square. Be sure, however, that the jambs are perfectly straight; otherwise the framing-square test will not be accurate.

10. Now finish nailing the exterior casings, placing the nails from 9 to 12 inches o.c. Nails should be so placed that they will be driven into a stud or header; this may require slanting them a little. One 16d finish nail is driven into each sill horn.

11. Remove temporary braces.

SETTING A WINDOW FRAME WHEN WATER TABLE IS USED. The same basic procedures are followed when setting a frame that has a subsill which fits over the lip on the water table, except that since the water table automatically levels the frame, no height rods are needed. Tack nails are driven into the lower end of the side casings, and the frame is checked for plumbness and levelness, the top ends are tack-nailed, and the frame is checked for squareness. It is then ready for final nailing.

*Setting a Doorframe.* The procedure used for setting a window frame is reversed in setting a doorframe; that is, checking is started from the bottom sill. The level part of the sill is placed so that it will be flush with the hardwood floor when that is laid. Proceed as follows:

1. Read the specifications to determine the thickness of the hardwood floor and make two small blocks to this exact thickness.

2. Set the doorframe into the opening and check the flat part of the sill for levelness with the blocks, which are placed on the subfloor to represent the hardwood floor. The recess made in the floor to receive the sill may need to be chiseled to get the frame sill to its correct height.

3. Tack-nail the lower end of the side casings.

4. Using a straightedge and level, plumb the two side jambs and tack-nail to place.

5. Check the frame for squareness, using a straightedge against the jamb while so doing.

6. Place wedges on the hinge side of the doorframe under the hinge-seat locations (usually 7 inches down from the head jamb to the hinge and 11 inches up from the bottom sill, with the third hinge, if one is used, centered between the top and bottom hinges). These wedges are

necessary to provide a solid support under each hinge and prevent jamb movement while the hinges are being cut in and screwed to place. Nail through the face of the jamb on through the wedges and into the trimmer studs, using 8d or 10d finish nails.

7. Place a wedge on the lock side of the frame, 36 inches from the floor to the center of the wedge. This wedge provides solid support under the lock striking plate. Nail as described above.

8. If necessary, wedge the jambs at any other place which shows a low spot on the straightedge.

**Note:** Wood shingles make excellent wedges for this purpose.

9. Face-nail the casings, using 8d box nails placed 9 to 12 inches on center.

10. Remove temporary braces.

11. Nail a 1-inch board on top of the sill to protect it during the construction period.


## TYPES OF ROOFING

Three types of roofing are used for residential buildings: wood shingles, composition shingles, and roll roofing. Roll roofing is best applied with hot tar. However, on small dwellings, sheds, farm buildings, etc., an inexperienced person can apply roll roofing without using hot tar, provided that he follows the proper instructions.

The discussion of the roofing unit is divided into four parts: (1) sheathing requirements, (2) wood shingles, (3) roll roofing, and (4) composition shingles.


## SHEATHING REQUIREMENTS

Sheathing for a wood-shingle roof is spaced so that the distance from the bottom edge of one board, usually a 1 by 4, to the bottom edge of the next board is twice the amount that a shingle is exposed to the weather, that is, twice the measurement of that part of the shingle which is seen. For a 4½-inch shingle exposure (see Figure 56), the sheathing is spaced 9 inches from edge to edge; for a 5-inch exposure the sheathing is spaced 10 inches.

Sheathing for composition shingles or for roll roofing is laid solid; 1 by 6 lumber, S1S1E, is used, the same as for subfloor.

For the cornice area of a roof (see Figures 21 and 22), the sheathing is always solid for the composition roofing and may be of tongue-and-

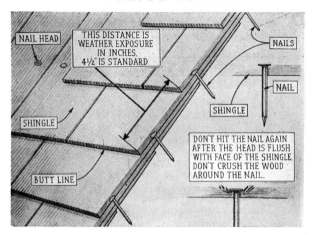

**Figure 56.** Shingle exposure.

groove finish lumber because the lower side is exposed. The T. and G. type of sheathing is often used on a cornice for a wood-shingle roof (see page 198).

## WOOD SHINGLES*

Wood shingles are made 16, 18, and 24 inches in length but are random widths, varying from 4 to 14 inches. The thickness is designated as 5/2, meaning that five shingles, when measured at the butt (thick end), will measure 2 inches. The best shingle is No. 1 red cedar; lower grades are known as No. 2 and No. 3. The basic difference in the grades is in the grain of the wood. No. 1 shingles are strictly vertical-grained.

Four bundles of shingles equal 1,000 shingles. One thousand shingles will lay one square (100 square feet of area).

Wood-shingle nails should be 3d box, 14½ gauge, galvanized, or 4d box, 14 gauge, galvanized. Roughly 3 pounds of nails should be allowed for 1,000 shingles, that is, for 100 square feet of area (see Table 1).

*Shingle-laying Rules.* There are several shingle-laying rules which must be observed if a watertight roof is to be obtained:

* For full details on the manufacture, seasoning, storage, installation, and staining of red-cedar shingles, see "Certigrade Handbook of Red Cedar Shingles," Red Cedar Shingle Bureau, Seattle, Washington, 1954. Illustrations 56 to 61, and Table 1, courtesy of the Shingle Bureau, acknowledgment of which is expressed with appreciation by the author.

**Table 1. Nail sizes for cedar shingles**

| FOR 16" AND 18" SHINGLES | | FOR 24" SHINGLES | FOR 16"& 18" SHINGLES | FOR 24" SHINGLES | FOR ALL SHINGLES |
|---|---|---|---|---|---|
| 1¼" LONG | 1¼" LONG *14½ GAUGE | 1½" LONG *14 GAUGE | 1¾" LONG *14 GAUGE | 2" LONG * 13 GAUGE | 1¾" LONG *14 GAUGE |
| APPROX.376 NAILS TO LB. | APPROX. 515 NAILS TO LB. | APPROX.382 NAILS TO LB | APPROX.310 NAILS TO LB | APPROX.220 NAILS TO LB | APPROX.380 NAILS TO LB. |

Square-cut nails of same length will also give satisfactory service. Standard "box" nails of the sizes given will prove satisfactory if properly zinc-coated or made rust-resistant.

SOURCE: Table courtesy Red Cedar Shingle Bureau, Seattle, Washington, "Certigrade Handbook of Red Cedar Shingles," 1954.

1. The shingles must be spaced with approximately ¼ inch between shingle edges. This allows for the expansion of the shingles when rain causes the wood to swell (see Figure 57).

2. When *one row of shingles is compared with the next row above,* side laps (see Figure 57) must be at least 1½ inches apart.

3. The first course of shingles at the cornice or edge must be double; otherwise the cornice sheathing would be exposed between the cracks (see Figure 21).

4. Two nails are required per shingle irrespective of its width, as succeeding rows of shingles provide additional nailing (see Figure 56).

5. The amount of shingle exposure from the lower end of one row to the lower end of the next row above varies according to the architect's requirements. The standard amount is 4½ inches, but a smaller exposure is used if certain architectural effects are desired; it is not advisable to make the exposure more than 5 inches. The smaller the exposure, the greater the number of shingles required (see Figure 56).

6. An experienced shingler will use a shingling hatchet with an ad-

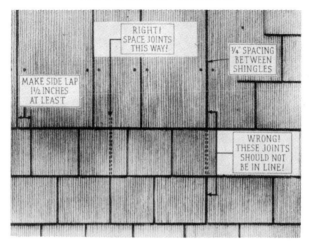

**Figure 57.** Spacing of wood shingles.

justable gauge, which is set to the required exposure. This gauge serves to locate each shingle in an exact position (see Figure 58). Occasionally he will strike a chalk line (see Figure 27, Chapter 6) to be sure the shingle butts are in a straight line.

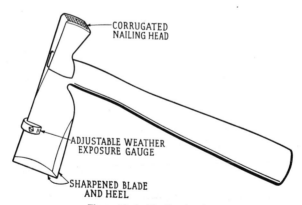

**Figure 58.** A shingling hatchet.

7. Valley tin (see Figure 59) is required at the intersection of two roofs. Valley tin, which is approximately 14 inches wide, has an inverted V in its center to prevent the water from running under the shingles which form the valley.

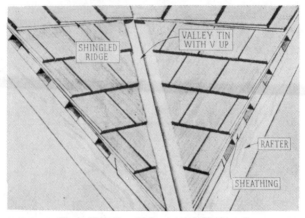

Figure 59. Valley tin for a wood-shingle roof.

8. Hips and ridges (see Figure 61, Chapter 6) are overlaid with a shingles ridge, with the lap, or overlay, alternated to avoid a continuous joint; this type is called the "Boston" hip (see Figure 60).

Figure 60. "Boston" shingled hip (ridge similar).

9. Nails should never be more than 1½ inches above the butt line of the next course (see Figure 56).

10. Care must be used to flash the shingles correctly against a chimney

face and end as illustrated in Figure 61. Galvanized iron should be used.

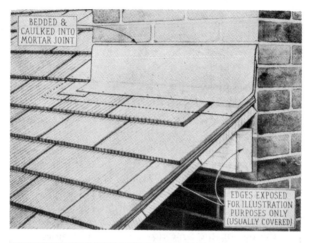

**Figure 61.** Chimney flashing.

11. A small seat is needed to prevent the shingler from sliding off the roof when the pitch is greater than results from a 3-inch rise and a 12-inch run. Make the seat as illustrated in Figure 62, using 1-inch material. To ensure that the seat will be level, see that the angle is cut correctly by using the steel square and marking and cutting on the run figure, 12 inches.

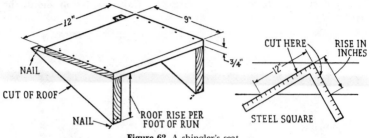

**Figure 62.** A shingler's seat.

12. The shingler should wear rubber-soled shoes (tennis shoes are excellent) to prevent slipping. On very steep roofs it is sometimes necessary to "shingle in" 1 by 3 pieces of lumber about every 18 or 20 rows. The cleats, which are about 12 feet long, are held in place by nailing the butt ends of several shingles flush with one edge of the 3-inch strip and on its lower side; the shingles that extend beyond the 1 by 3 are then nailed onto the roof in the same manner as for any shingle. This will provide a 1 by 3 the full length of the roof. After the roof is completed, the cleats are removed by sawing off the shingles to make them even with the shingle butts in that row (see Figure 63).

A strong ladder must be used when the bundles of shingles are carried to the roof. This ladder should be carefully inspected for weak or damaged sides or treads before the shingler puts his weight on it. With the above information in mind, proceed as follows when shingling a roof with wood shingles:

### Detailed Instructions:

1. Carry sufficient bundles of shingles to the roof to cover the area involved (four bundles for every 100 square feet), distributing them over the area and keeping them well above the lower portion of the roof to provide work clearance.

A bundle will remain in place in a horizontal position if one end is pushed between the sheathing boards so that a few shingles catch in this space. Bundles are *never* opened until they are ready for use; otherwise they would blow or slide all over the roof.

2. Nail on the first shingle at one end of the roof, keeping the lower end (or butt) 1½ inches below the edge of the cornice edge of the first sheathing board (see Figure 64). To prevent the shingle nails from protruding through the cornice (succeeding rows of shingles are double—hence the nails will not go through the cornice sheathing), drive only

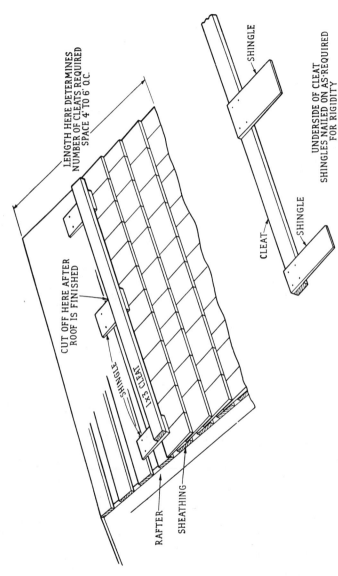

SHINGLE

UNDERSIDE OF CLEAT
SHINGLES NAILED ON AS REQUIRED
FOR RIGIDITY

CLEAT

SHINGLE

LENGTH HERE DETERMINES
NUMBER OF CLEATS REQUIRED
SPACE 4' TO 6' O.C.

CUT OFF HERE AFTER
ROOF IS FINISHED

SHINGLE

1 X 3 CLEAT

RAFTER

SHEATHING

**Figure 63.** Shingling in a supporting cleat while shingling a steep roof.

about two-thirds of their length into the sheathing and bend the nails flat.

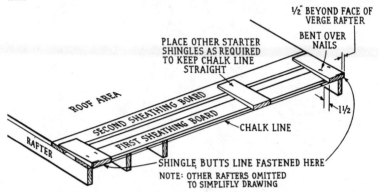

**Figure 64.** Starter shingles.

3. Nail on the second shingle, in the same manner, at the other end of the roof.

4. Stretch a chalk line tightly from the first shingle butt to the second shingle butt to serve as a guide line for all shingles in the first row.

5. Nail on several other shingles at intermediate points between the two end shingles to hold the chalk line in an exact position. These shingles can be nailed on almost anywhere since it is easy to fit shingles to them by either selecting narrow ones or splitting a wide one to the required width.

**Note:** It may be more convenient to do steps 2 and 5 from a ladder.

6. The shingling job is now correctly started. The two end shingles are in place, and a chalk line is in position to guide the placing of the first row of shingles which project beyond the cornice sheathing board. After the first row is nailed onto this line, it must be covered (doubled) with a second row of shingles nailed directly over the first row. This is necessary to cover the sheathing which is exposed between the edges of the shingles in the first row. Allow $\frac{1}{4}$ inch between each shingle and at least a $1\frac{1}{2}$-inch side lap at every joint (see Figure 57).

7. Set the gauge of the shingling hatchet, which is used to great advantage for the remainder of the job, to the required shingle exposure ($4\frac{1}{2}$ inches or 5 inches, etc.).

8. Each shingle that is placed on the roof is located by means of two measuring points (see Figure 65): (1) the lower front corner (butt edge) of a shingle just nailed, which serves as the guide to locate the lower back corner of the next shingle; and (2) the shingling hatchet, which

locates the front or "free" edge of the new shingle to be nailed on. Hold the gauge against the butt edges of the last row of shingles nailed to place and set the new shingle butt exactly against the head of the shingling hatchet. Never place the gauge in the center of a shingle; the lower edges of the shingle rows will soon become crooked, and each succeeding row will become continually worse.

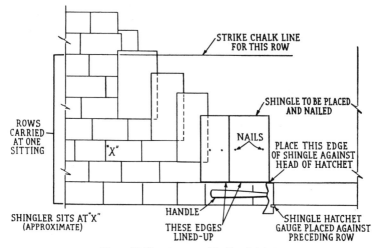

**Figure 65.** How to use a shingling hatchet.

9. To be sure that the rows of shingles do remain straight, strike a chalk line from one end of the roof to the other for at least every seventh row; then lay a row of shingles exactly to this line.

10. Continue in this manner. A little experience will show how many shingle rows can be nailed on during one move across the roof (see Figure 65). Usually five rows is the limit of reach from the shingling seat. An expert may carry seven rows. If too many are attempted, a crooked line of butts will result.

11. After a few rows have been completed, measure the amount of exposure at each end of the last row of shingles. Strike a chalk line and hold a row exactly to the line; this will automatically correct any discrepancies in the line of shingle butts.

12. When within 4 or 5 feet of the roof ridge, measure the distance, at each end of the roof, from the last row of shingles to the ridge. Failure to make the last row parallel to the ridge is considered poor shingling. To adjust any difference, shorten the spacing at the narrow end (or widen the space at the wide end), strike a line, and proceed as before. A slight

change in the spacing of the shingles, say, ¼ or ⅜ inch, is not apparent to the eye and will soon "take up" any difference in the measurements to the ridge.

13. Flashing against a chimney, as illustrated in Figure 61, should be done as the work progresses. Be sure that the flashing enters the brick mortar joint and that the joints are well caulked with a caulking compound after the metal is in place.

**Note:** To cut the mortar to receive the metal flashing, a mortar-joint "saw" can be easily made from a piece of heavy sheet metal, say 24- or 26-gauge. Using a pair of tin snips, cut a set of teeth, about ½ by ½ inch, into the edge of a piece of metal approximately 12 inches long and 4 inches wide. Nail the metal onto a piece of wood to serve as a handle and use as you would an ordinary saw. The teeth will need recutting occasionally (see Figure 66).

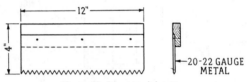

**Figure 66.** A mortar-joint saw.

14. Valley tin is placed as shown in Figure 59. This can be bought at a sheet-metal shop; it should not be less than 14 inches wide for steep roofs and 10 inches wide for a flatter roof.

The sheet-metal man, if he is given the roof pitch and the length of the valley, will bend the metal to an exact angle and solder the ends together to the correct length. The center inverted V (see Figure 59) is all-important to prevent water from seeping under the shingles when rain gives a heavy runoff into the valley.

The metal should be painted with red lead prior to placing it into position, to prevent rusting and corrosion.

The lower end is placed so that the center of the valley tin projects at least 1 inch beyond the cornice line. Excess tin can be cut off to suit the cornice shape.

Only a few nails along each edge are required to hold the valley in place prior to shingling. Use shingle nails and drive them into the roof sheathing.

To shingle into a valley, proceed as follows:

*a.* Strike two lines from the top end to the bottom end of the valley tin to serve as a guideline for the shingle edges (see Figure 67*A*). It is customary to leave a 4-inch space between the chalk lines at the top and a 6-inch space between the shingles at the bottom. This gives a "funnel" effect and prevents water from backing up in a heavy storm.

b. Cut one shingle to use as a pattern for sawing the shingles required to fit into the valley. This is done on the roof and made to fit the job conditions. (The pitch of the roof changes the angle of the shingle edge. See Figure 67B.) If the roof pitch is different on one side of the valley, two patterns will be required. Once the pattern (or patterns) is made, the work of preparing the valley shingles can be done on the ground.

c. Select several good shingles of even width—say 6 or 8 inches—stack them, and fasten them together, nailing through the thin or top ends (see Figure 67C). Mark the lower end from the pattern and cut, using a fine-tooth saw. This procedure will ensure that the completed shingled valley will have a straight edge.

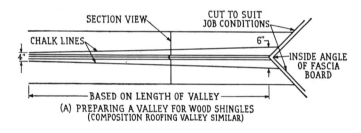

SECTION VIEW
CUT TO SUIT JOB CONDITIONS
CHALK LINES
6"
4"
INSIDE ANGLE OF FASCIA BOARD
BASED ON LENGTH OF VALLEY
(A) PREPARING A VALLEY FOR WOOD SHINGLES
(COMPOSITION ROOFING VALLEY SIMILAR)

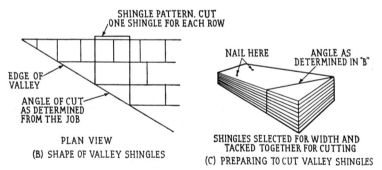

SHINGLE PATTERN. CUT ONE SHINGLE FOR EACH ROW

NAIL HERE
ANGLE AS DETERMINED IN "B"

EDGE OF VALLEY

ANGLE OF CUT AS DETERMINED FROM THE JOB

PLAN VIEW
(B) SHAPE OF VALLEY SHINGLES

SHINGLES SELECTED FOR WIDTH AND TACKED TOGETHER FOR CUTTING
(C) PREPARING TO CUT VALLEY SHINGLES

**Figure 67.** Preparing a composition valley for shingling.

15. Shingle hip rafters in the same manner as the valleys: prepare the shingles in advance by making a pattern to suit the roof pitch, nailing together several shingles of equal width, and cutting them with a fine-tooth saw.

16. Shingle ridges and hips as illustrated in Figure 60. Prior to

shingling, select shingles that are 4 or 5 inches in width from the various bundles. Lay them as shown, alternating the laps and using the blade of the shingling hatchet to shave off the overwood to conform to the face of the opposing shingle.

17. The last step is to "broom-clean" the roof. This should be done cautiously, for stepping on a piece of loose wood may easily result in a bad fall. The value of shoes with rubber soles will be most apparent in this operation.

## SINGLE-LAYER ROLL ROOFING*

*General Information.* A roof to be covered with single-layer roofing must be solid-sheathed. All large knotholes in the sheathing should be covered with tin. This is very important; otherwise a shoe heel might easily punch a hole through the paper. Using scrap tin and a pair of tin snips, cut the tin a little larger than the knot to be covered. Use 3d shingle nails to tack each piece in place.

Each roll of roofing is 36 inches wide and 36 feet long, containing 108 square feet of material. One roll covers 100 square feet of area; the additional 8 square feet are used for the side laps (at least 2 inches) and end laps (at least 4 inches).

The roofing is manufactured by weight per 100 square feet of coverage (one roll). A standard roll, which actually weighs 87 pounds, is known in the trade as 90-pound roofing. Each roll is packaged with a can of roofing cement and with a sufficient quantity of flat-headed $7/8$-inch nails to fasten it to the roof.

Single-layer roll roofing is laid up or down or horizontally, depending on the pitch of the roof. Roofing manufacturers recommend horizontal laying for roofs of more than 7-and-12 pitch.

**Note:** Laying the roof vertically and not horizontally enables the roofer to do much of his work on top of the sheathing instead of on top of the composition roofing (until the last piece of roofing paper is laid). This procedure not only protects the finished roof but makes it easier on the roofer, for in very warm weather composition roofing becomes almost too hot to walk on.

Composition roofing paper should be unrolled, cut to length, and exposed to the sun prior to nailing it to the roof. It should be laid during mild weather, for laying it during cold weather may cause wrinkles to appear afterwards. For small buildings the paper can be cut on the ground, provided that a suitable flat area, free from dirt, can be found.

* The following information is adapted from material furnished by Pabco Products, Inc., San Francisco, Calif. Permission to use is acknowledged with appreciation by the author.

**Note:** Roofing-paper strips should never be laid flat on a lawn while the roofing work is proceeding, for if they are left for any length of time, the hot sun will burn the grass under them.

The best tool to use for cutting roofing paper is a roofing knife, which has a hooked end with a sharp point (see Figure 68). The roofing cement must also be warmed up in the sun. A 2-inch paintbrush is best for applying the cement to the roofing paper.

FLASHING A VALLEY. To flash a valley, strips of roofing paper must be cut 12 and 24 inches wide. The paper is first cut to the required hip or valley length; then cut lengthwise to make the 12- and 24-inch strips.

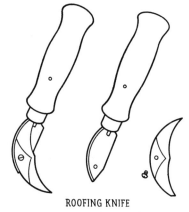

ROOFING KNIFE

**Figure 68.** A composition-roofing knife.

To flash a valley correctly, proceed as follows (see Figure 69):

1. Lay a 12-inch-wide strip of roll roofing, as shown at *A*. When using mineral-surfaced roofing, lay mineral side down.

2. Nail the roofing to place, spacing the nails 12 inches apart and 1 inch from edges, as at *B*.

**Figure 69.** Flashing a valley for composition roofing.

3. Apply a strip of cement 2 inches wide to both edges of the 12-inch strip (*C*).

4. Lay a 24-inch-wide strip of roll roofing on top of and in the center of the 12-inch strip and press it into the cement (*D*).

5. Nail the roofing to place, spacing the nails 12 inches apart and 1 inch from edges (*E*).

Make two chalk lines to serve as a guide for the ends of the roofing paper. Leave a 6-inch space between the lines at the top and increase it in width ⅛ inch for every foot of valley (see Figure 67). The ends of the roofing strips are cut to fit the angle and laid to the chalk lines.

HIP AND RIDGE FLASHING. Hip or ridge flashing is done *after* the roofing paper is laid. Proceed as follows (see Figure 70):

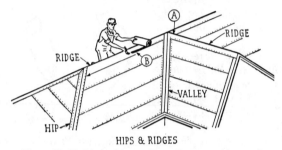

**Figure 70.** Flashing a hip or ridge of a composition roof.

1. Cut 12-inch-wide strips of roll roofing, as shown at *A*.

2. Apply 2-inch-wide strips of roofing cement along the under edges of the 12-inch-wide strips. Place strips in position, press down, and nail along each edge, spacing nails every 2 inches, 1 inch in from edge, as at *B*.

CHIMNEY AND FIRE-WALL FLASHING. Chimneys are metal-flashed as illustrated for wood shingles (see Figure 61); the metal flashing is placed under the roofing as the work progresses. For composition roofing, how-

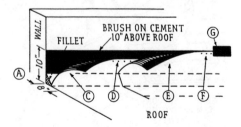

**Figure 71.** Flashing fire wall for a composition roof.

ever, a 4 by 4 wood angle fillet is required at the junction of the roof and fire wall or chimney. Flashing can also be done with composition roofing materials. Proceed as follows (see Figure 71):

1. Nail a 4 by 4 wood angle fillet (see *A*) the full length of the junction of roof and wall.

2. Prime the wall 12 inches above the roof line with roofing cement and allow to dry.

3. Turn up the roofing 6 inches against the wall and cement it in place with Hydroseal, or an equivalent (*C*).

4. Nail the roofing 1 inch on centers (*D*).

5. Trowel a ¼-inch layer of Hydroseal on a strip of roofing that is 18 inches wide and the same length as the wall. Then bend the strip lengthwise at a 90-degree angle, cementing 10 inches of the width dimension to the roof and 8 inches to the wall (*E*).

6. Nail roofing at upper edge with large-headed galvanized roofing nails (*F*).

7. Trowel a ¼-inch-thick strip of Hydroseal over nails and 2 inches up the wall (*G*).

CORNICE EDGES. The cornice edges at the gutter line and at the verge-rafter edge (see Figure 72) are finished by bending the paper ¾ inch over the edges of the roof, nailing it to place, and covering it with a strip of 1 by 2 S4S finish lumber.

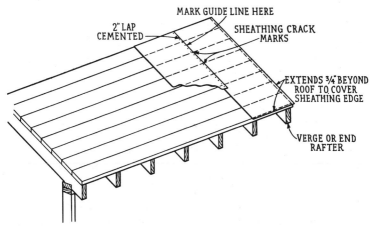

**Figure 72.** Locating sheathing edges when laying composition roofing.

*Procedure for Laying Roll Roofing.* With the above general information in mind, proceed as follows when laying a composition roof.

Cut the roofing into strips of the correct length and place them where they can warm up, either on the roof or on the ground. Carry the rolls of paper to the roof and place them within easy reach, but do not pile them together. Wear rubber-soled shoes to prevent the feet from slipping, sweep the roof area clean and cover large knots and knotholes with tin. Measure the roof from the cornice edge to the ridge. On small buildings measure from one cornice edge to the cornice edge on the opposite side of the roof. The procedure to be followed at this point will depend upon whether the roofing is to be laid according to the up-and-down method or the horizontal method.

UP-AND-DOWN ROOFING:

1. Lay the first strip parallel to the end of the roof, letting it extend ¾ inch for the turndown over the verge rafter.

2. Drive two roofing nails near the top and close to the edges of the strip. This will hold the paper in its correct position.

3. Using the weight of the body to help stretch the paper, slide down on top of the paper and drive several nails on the inside edge. Nails are needed only about every 6 feet. Be sure the paper remains parallel to the verge rafter.

4. On the verge-rafter edge bend the paper over the corner and nail every 2 inches, keeping the body weight above the nails. If roofing strip has been warmed up it will bend easily.

5. On the "inside" edge, using a pencil or lumber crayon, mark the location of each sheathing-board crack. Marks must be kept in a little more than 2 inches because the next roofing strip will lap 2 inches over the first one. Marks should be made lightly so that they will not show after the job is completed (see Figure 72).

6. Repeat this process for the second strip. Be sure that it is placed exactly parallel to the first one by measuring in 2 inches at the top and bottom ends of the first piece and then placing the second strip exactly even with these two marks. Tack-nail the inside or "free" edge of the strip. Then make a pencil line the full length of the lap.

7. Raise up the lapped edge and brush on a coat of warm roofing cement, keeping it between the pencil line and the edge of the roofing; otherwise an unsightly seam will result.

8. Carefully "walk down the seam," pressing the two pieces of roofing paper together with the toes of your shoes. Then, beginning at the top, nail down the seam, spacing the nails 2 inches apart. The pencil lines representing the sheathing cracks will guide the nailing so that the nails will not be driven into a crack.

9. Continue in this manner. If the last part of a roll is not long enough for a full-length roofing strip, let the horizontal seam come where it may and finish with a second piece. Be sure that the upper piece of

roofing laps at least 4 inches over the lower piece. This horizontal lap should not be made too near the edge of a sheathing board.

HORIZONTAL ROOFING:

1. Measure the roof length and cut several strips of roofing paper, allowing ¾ inch for the turndown at each gable end.

2. Lay a 36-inch-wide roofing strip so that the bottom edge and ends extend ¾ inch over the eave and rake, and nail the top edge every 18 inches, keeping the nails in 1 inch from the edge, as shown in Figure 73. This will hold the sheet in exact position and give it an opportunity to expand.

EAVES & EDGES

**Figure 73.** Starting first strip of a composition roof laid horizontally.

3. Nail down the ends and lower edge, spacing the nails 2 inches apart. This job will have to be done from a ladder.

4. Repeat this process for the next strip, being sure to allow for a 2-inch lap. After the top edge is nailed, make a pencil line along the lower edge to serve as a guideline when brushing on the cement.

5. Lift up the lower edge, brush on a good coat of roofing cement, press firmly into place, and nail every 2 inches (see Figure 74).

**Note:** Be sure that the nailing for the lower edge of the roofing paper does not come too close to the edge of a sheathing board. The strip may have to be lowered slightly to avoid the crack.

6. Continue this procedure until you reach the ridge; then repeat for other roof surfaces.

**Figure 74.** Cementing strips for horizontal laying of composition roofing.

After completing the procedure described for either up-and-down or horizontal roofing, cap hips and ridge with 12-inch strips of roofing material (see Figure 70). Waste remaining from the valley or hip flashing can be utilized, as end laps (4 inches) can be made anywhere. Then cover the eave and rake edges with a 1 by 2 S4S finish strip of lumber. Gable ends can be done from the roof. Eave strips at the cornice edge require the use of a ladder. Finally, broom-clean the roof, being careful to remove any loose nails.


## COMPOSITION SHINGLES

*Manufacture, Shape, and Size.* Composition shingles are manufactured in a variety of shapes and colors. They are made as single pieces or with three shingles in one strip. Three bundles will provide coverage for 100 square feet.

Local dealers should be consulted about selecting a shingle that conforms to the architecture and general appearance of the house. Too large a composition shingle may spoil the appearance of a well-designed house. The color should harmonize with the general color scheme; sometimes the color of the body of the house is not determined until after the composition shingle is selected, for outside wall paint can easily be changed to blend in with the roof colors, whereas the color selection for shingles is more limited.

*Footrests.* For steep roofs footrests, made of 2 by 3 or 2 by 4 stock, are required to prevent the roofer from sliding off the roof. These may be held in place either by ropes attached to hooks that lock over the ridge or by 2 by 3s placed vertically and hooked over the ridge, as shown in Figure 75. The 2 by 3s must be prevented from cutting or chafing the shingles.

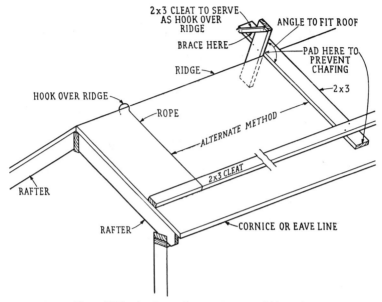

**Figure 75.** Staging for roofing a steep composition roof.

***Nailing.*** Use galvanized $10\frac{1}{2}$- to 12-gauge nails with minimum $\frac{3}{8}$-inch heads. When laying composition shingles over sheathing boards, use nails $1\frac{1}{4}$ inches long. When laying over old wooden shingles, use nails $1\frac{1}{2}$ inches long.

A strip of galvanized iron or well-seasoned and painted wood should be nailed along the otherwise unprotected side edges of the roof.

***Laying Composition Shingles.*** When composition shingles are laid, valley rafters should be flashed as shown in Figure 69, hips and ridges as shown in Figure 70, and chimneys as shown in Figure 61. To lay the shingles, proceed as follows (see Figure 76):

1. Place a 12-inch strip of mineral-surfaced roll roofing along the eaves as a starting strip. Use only a few nails to hold this strip in place. In snow country the starting strip should be wide enough to come higher than any possible backup of water caused by snow at the eaves.

2. Invert and lay the first course of shingle strips along the eave, placing the tabs at the top. A tab is that part of the shingle exposed to the weather.

3. Lay the next course with the tabs coming out to the eave.

4. The shingle strips of each succeeding course should be laid to break

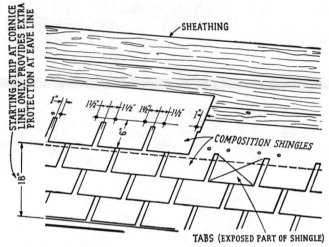

STARTING STRIP AT CORNICE LINE ONLY PROVIDES EXTRA PROTECTION AT EAVE LINE

SHEATHING

COMPOSITION SHINGLES

18"

1"   1½"   1½"   1½"   1½"   1"

6"

TABS (EXPOSED PART OF SHINGLE)

**Figure 76.** Laying composition shingles.

joints with the course below; that is, strip ends should always be 2 or more inches out of direct line with the notches in the course above and below. If desired, the shingle strips may be laid regularly, the centers of the tabs exactly corresponding to the notches between the tabs of the shingles in the course below and above. Each course should be set back to expose the whole of the tabs in the course below.

Nail as shown in Figure 76. Each nail should be covered by the tab on the shingle strip above, which should extend 1 inch beyond the nail.

**Note:** Drive nails squarely with the roof surface to avoid cutting the shingles with one edge of the nailhead.

5. Broom-clean the roof, being sure also to clean out the gutters.

# CHAPTER EIGHT

# Interior Finish

The interior-finish work of a house can, for discussion purposes, be conveniently divided into six divisions or units as follows: (1) the various parts which make up the interior finish of a house; (2) making and setting doorjambs; (3) fitting and nailing wall trim; (4) fitting and hanging doors, windows, and sash; (5) making cabinets; and (6) fitting, nailing, and scraping hardwood floors. Each of these units will be illustrated and described in detail.

## INTERIOR-FINISH PARTS

The different parts of a house which are classified as "interior finish" are, for the most part, bought as milled material from a lumberyard or planing mill. These parts include casings, baseboard, and picture molding, as illustrated in Figure 1; window stool and apron, as shown in Figure 2; and hook strips and clothes-pole rosettes, as illustrated in Figure 3. Figure 4 indicates two typical doorjamb shapes and also includes doorstops and threshold.

**Note:** Tongue-and-groove flooring, as shown in Figure 1, is often used to make a flush baseboard. It is fitted and nailed to the studs *before* the walls are plastered. The groove is placed up to make a good plaster key. The flush base eliminates the dust-catching edge found when the baseboard is applied *after* the plastering is completed.

Figure 5 shows some typical interior doors, such as a one-panel door, a three-panel door, a detail door (so called because this type of design is usually manufactured "to detail"), a 10-light French door, and a flush or slab door.

A two-light window and a top-four-light double-hung window are

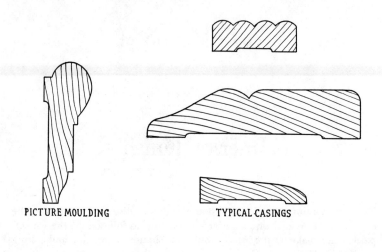

PICTURE MOULDING         TYPICAL CASINGS

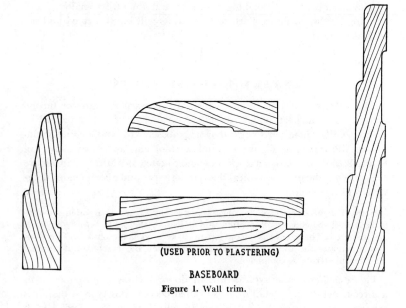

(USED PRIOR TO PLASTERING)

BASEBOARD

**Figure 1.** Wall trim.

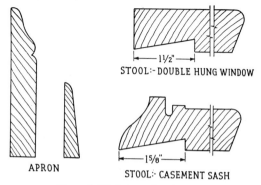

STOOL:- DOUBLE HUNG WINDOW

— 1½" —

— 1⅝" —

APRON

STOOL:- CASEMENT SASH

**Figure 2.** Window stool and apron.

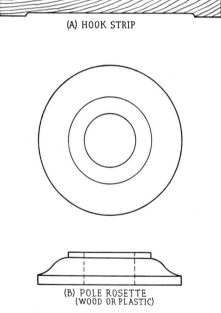

(A) HOOK STRIP

(B) POLE ROSETTE
(WOOD OR PLASTIC)

**Figure 3.** Hook strip and pole rosette.

THRESHOLD

INTERIOR DOOR STOPS

PLAIN JAMB

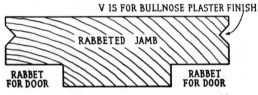

V IS FOR BULLNOSE PLASTER FINISH

RABBETED JAMB

RABBET FOR DOOR                    RABBET FOR DOOR

**Figure 4.** Doorjambs, doorstops, and threshold.

shown in Figure 6, as well as a one-light casement sash, a 12-light case-ment sash, and a pair of 12-light casement sash.

The cabinets found in an average home include a sink case, dish cup-board, broom closet, soap cabinet, book cabinet, linen case, ironing board (not exactly a cabinet but classified as such), and a wardrobe cabinet. Some of these cabinets are illustrated in Figure 7.

Figure 8 illustrates a typical mantel shelf.

With the exception of the doors, windows, and cabinets, the millwork items listed and illustrated are usually purchased in long lengths and then cut and fitted on the job to their respective places. Doors, windows, and sash are ordered as individual items.

Cabinets can be purchased (1) as stock items, in which case the house frame must be exactly laid out to receive ready-made cabinets; (2) as "tailor-made" items requiring the millman to go to the house under construction, take actual job measurements, and build the cabinets to these measurements.

A third method or procedure is to build the cabinets "on the job." Selected finish and shelving stock, say 1 by 10 or 1 by 12 S4S, is ordered to make the skeleton frame or "carcase." This stock is cut to length and glued to make jambs, partitions, and shelves of the required width.

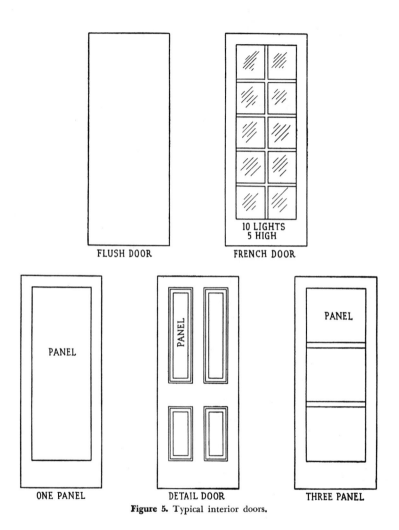

FLUSH DOOR  FRENCH DOOR

10 LIGHTS
5 HIGH

PANEL

PANEL

PANEL

ONE PANEL  DETAIL DOOR  THREE PANEL

**Figure 5.** Typical interior doors.

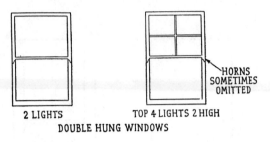

2 LIGHTS          TOP 4 LIGHTS 2 HIGH

HORNS
SOMETIMES
OMITTED

DOUBLE HUNG WINDOWS

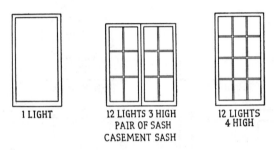

1 LIGHT          12 LIGHTS 3 HIGH          12 LIGHTS
                 PAIR OF SASH             4 HIGH
                 CASEMENT SASH

**Figure 6.** Typical windows and sash.

Front materials and drawer stock are ordered in long lengths as milled items, then cut as needed to make a specific part of a cabinet.

*Cutting Finish Stock to Rough Lengths.** Long lengths of finish lumber must first be cut into rough short lengths so that the various pieces of casings, apron, stools, baseboard, hookstrips, etc., can be handled easily when they are fitted to their places. Rough-length measuring must be carefully done to avoid waste or a shortage of finish stock.

When it is time to start work on the interior finish of a house, it is customary for one carpenter to precede a crew of finish carpenters, a day or so in advance, to do rough trim cutting and, after cutting, to place the various pieces of finish, including doors and windows, in each room as required. This is good trade practice as it places the responsibility for cutting the trim economically on one carpenter. It is then possible for a carpenter to complete the trimming of a room without having to look

* For detailed information on how to make a cutting list of various interior-trim items, see J. Douglas Wilson and Clell M. Rogers, "Simplified Carpentry Estimating," Simmons-Boardman Publishing Co., New York, 1956, chap. 5.

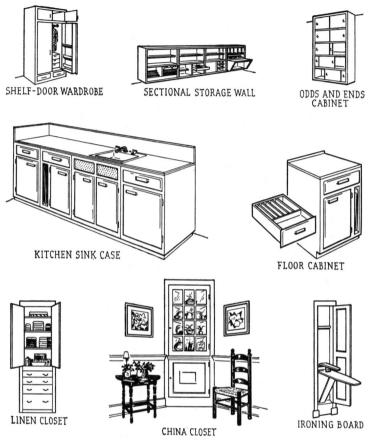

SHELF-DOOR WARDROBE    SECTIONAL STORAGE WALL    ODDS AND ENDS CABINET

KITCHEN SINK CASE

FLOOR CABINET

LINEN CLOSET    CHINA CLOSET    IRONING BOARD

**Figure 7.** Typical cabinets.

for any finish materials. (In mass housing a carpenter does only one job, repetitively. Procedures outlined in this book pertain to the all-round carpenter.)

The best finish stock is selected for the most prominent places, such as the living room, dining room, or den. Slightly defective pieces or cross-grained lumber should, whenever possible, be used in the closets or in other places where they may not be easily seen.

Casings that are to be scribed against a wall (requiring that they be ripped) should be carefully selected for "softness" so that the ripping and nailing job will be as easy as possible.

A doorjamb requires four side casings, two head casings, two side stops, and one head stop; a window frame requires two side casings, one head casing, one piece each of stool and apron, and—if it is for a double-hung window—two side stops and one head stop. Lengths of door and window trim are based on the dimensions of the doors and windows.

Baseboard must be cut to rough lengths corresponding to the wall spaces in the various rooms. Measure each space, allow 1 to 2 inches for finish cutting, and write down the result. Keep the list for each room separate so that each piece of baseboard can be properly marked on the back to identify its correct location. Then, in order to utilize all the stock, measure each long piece of baseboard, cut the *longest lengths first,* and work down the list to the shortest pieces.

Hookstrips, clothes poles, and closet shelving are rough-cut in a similar manner.

This rough-cutting procedure results in keeping the floor clear of lumber, thus making it easier to use the workbench and to broom-clean the house occasionally to remove the scraps of lumber and piles of shavings which are constantly accumulating. This clean-up procedure is also an excellent fire-prevention measure.

MANTEL, EARLY AMERICAN

**Figure 8.** Mantel shelf.

*Tool Usage.* Space does not permit listing the various tools necessary to do interior-finish work (see Chapter 3). Suffice it to say that fine craftsmanship is obtained only by having the proper tools to do the various types of fitting, cutting, and assembling required in doing interior-finish work of the scope described in this chapter. A kit of *sharp* tools is also essential to do expert finish work.

*Jigs to Facilitate Finish Work.* There are several jigs, made of wood, which skilled carpenters have found almost indispensable when doing finish work (see Figure 9). They are easily made; the gauge block and the "preacher" should be kept in the finishing-tool box in readiness for use when needed.

STEP BENCH. (See Figure 9A.) A strongly made step bench is a necessity in trimming doorjambs; the only alternative is a sawhorse, which is usually too high to be used conveniently. The step bench is made by nailing two 12-inch legs of 1 by 12 material onto a 1 by 12 top, 18 inches

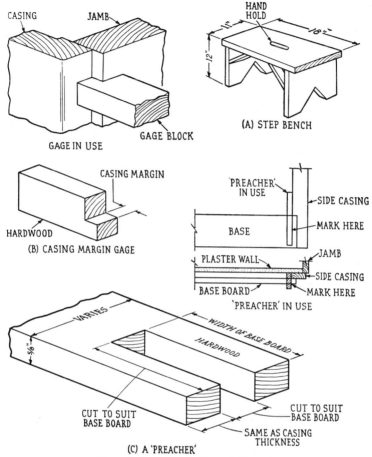

Figure 9. Jigs which facilitate finish work.

long. Legs are kept close to the ends of the top (to prevent tipping) and must be well braced. A hole, 1 by 4 inches, is cut in the center of the top to make the step bench easy to handle.

CASING-MARGIN GAUGE. (See Figure 9B.) Sometimes, in lieu of a gauge line on the edge of the jamb, a carpenter makes a casing-margin gauge block. The block is made from a small piece of hardwood, ¾ by ¾ by 2 inches in size. A ¼-inch rabbet, the exact size of the margin, is cut on one end. The square shoulder of the rabbet is placed against the jamb, and the casing is then placed against the end of the block and nailed. The block is moved as the nailing progresses.

PREACHER. (See Figure 9*C*.) A "preacher" is simply a piece of hardwood of suitable size that has been slotted to slip over a piece of baseboard when the base is fitted to a side casing of a doorjamb. The slot is made to conform to the size of the base.

Note: There is no known explanation for the term *preacher;* the word, however, is in common use among experienced carpenters.

*Nailing Information.* The size of nail to use for interior-trim items varies with the thickness of the lumber. A casing nail is often used; this nail is slightly larger than a finish nail and has a tapered head, which spreads the wood as it is driven in; a finish nail has a square shoulder and, in effect, breaks the grain of the wood as it is driven in.

Note: To avoid splitting the wood when nailing near the end of a piece of finish material, place the nailhead on a piece of metal (a flat-headed nail on the workbench will do) and blunt the nail point slightly. The nail will then act as a punch rather than as a wedge; the blunt end will "push" the wood ahead of itself as it is driven in and thus keep the wood from splitting. This procedure will not work for hardwood: in this case, drill a small "pilot hole," the same diameter as the nail.

All finish nailheads must be concealed by the painter as he does the painting and decorating of the interior trim and woodwork; therefore, all finish nails which are exposed must be set. To avoid missing any, set each nail as it is driven. Nail sets of several different sizes are required to fit the various sizes of finish nails.

*Millwork Delivery Suggestions.* Manufacturers of millwork recognize the importance of keeping such millwork items as doors and sash and interior trim perfectly dry, and all good millwork is made from kiln-dried lumber. The absorption of moisture can cause the warping and twisting of milled items, the raising of the grain, and ultimately the opening up of joints. The following suggestions, therefore, should be given careful consideration:

1. Do not have doors or other mill products delivered or stored on the job until the plaster is dry. If possible store them in a closed, dry room until they are primed.

2. Prime mill products with one coat of thin paint as soon as possible.

3. Give top and bottom edges of doors two coats of paint or white lead as soon as they are fitted to openings.

4. Be sure that the drying-out process in the building is complete *before installing* the mill products.

## DOORJAMBS

The term *jamb* has several meanings: (1) the finish frame into which an outside door is hung; (2) jamb stock, meaning lumber milled specifically for that purpose; and (3) one piece of a jamb, technically called *side jamb* or *head jamb*. On the job these terms are used rather loosely.

*Thickness of Jamb Stock.* The thickness of jamb stock depends on the finish around the jamb. One-inch stock ($\frac{13}{16}$ inch net) is standard when wood casings are used (see Figure 10A). One-inch-net thickness (requiring $1\frac{1}{4}$-inch stock) is necessary when the plaster is to be bullnosed around the jamb. Two-inch stock, finishing $1\frac{5}{8}$ inch, is required to make a rabbeted jamb (see Figure 10B).

*Width of Jamb Stock.* The width of the jamb stock depends on the type of finish required. When casings are used, the jamb is made the same width as the exact thickness of the wall including the plaster (see Figure 10A). If the job calls for a bullnosed plaster finish, the jamb stock is made the exact width of the wall studding. A V is milled into the edges of the jamb to receive the plaster, in order to make a tight joint between the jamb and the plaster (see Figure 10B).

Sometimes a molded-edge jamb is used; in this case the stock is made $\frac{1}{2}$ inch wider than the plaster-to-plaster measurement, and a V is milled into the back side of the jamb to receive the plaster (see Figure 10C).

*Joints.* A doorjamb consists of two sidepieces and one headpiece. When casings are used, the joints at the corners are dadoed together. The dado can be cut into either the head jamb or the side jamb, as illustrated in Figure 10D. The depth of the dado is usually $\frac{3}{8}$ inch.

For a plain jamb with a bullnosed plaster finish the rabbet joint is best for joining the head jamb to the side jambs (see Figure 10B). If the jamb is rabbeted, a combination of miter and rabbet is used, as shown in Figure 10B.

When the jamb is molded, the corners are mitered and rabbeted to assure perfect alignment of all arrises of the molding (see Figure 10C).

*Length of a Doorjamb.* The length of the jamb from the *underside of the headpiece to the bottom end* of the jamb is made the same as the door which is to be fitted into the finished jamb, *plus* $\frac{1}{2}$ inch for rug

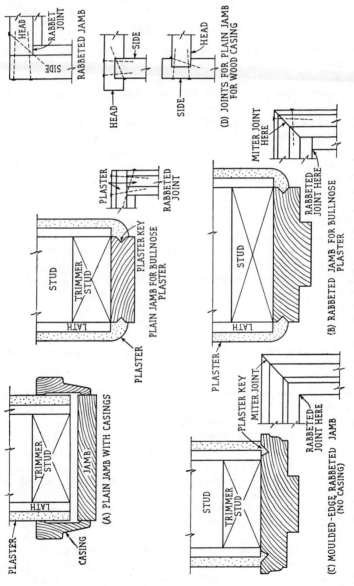

**Figure 10.** Optional methods of making an interior doorjamb.

clearance. For instance, a 6'8" door requires a 6'8½" doorjamb measurement.

*Width of a Doorjamb.* The width of the finished doorjamb is made the same as the width of the door, although it is customary to measure the doors, when possible, and make the doorjamb slightly wider to eliminate some planing when the door is fitted.

*Making a Doorjamb.* A list of the doorjambs to be constructed should be made before starting any layout work. In the following description it is assumed that the jamb stock has been rough-cut into sides and heads (as outlined previously).

The doors in an average house will vary in width, but 2'6" and 2'8" are the standard dimensions. The door lengths most commonly used are 6'6" and 6'8".

The following steps describe how to make one plain doorjamb.

1. Read the plans to determine the sizes of the doorjambs.

2. If the doors have been delivered to the job, measure their widths to determine if they have been made net or slightly oversize; also inspect the door edges to be sure that they are in good condition. Occasionally, through rough handling, a door may become bruised, necessitating considerable edge planing to remove the blemishes. (In this case the jamb width should be made slightly undersize.)

3. Read the detail to determine whether the jamb stock is to be plain or rabbeted and whether the plaster is to be bullnosed or wood casings are required. Select two side jambs and one head jamb.

4. Plane one edge of all pieces straight and slightly beveled (unless the jamb has molded edges).

5. Measure the wall to determine the finished width of the jamb desired (see parts *A* and *B* of Figure 10).

6. Gauge from the straightened edge, and plane slightly "under" to width. Jambs should be slightly wider than the thickness of the finished wall.

7. Determine whether the head jamb or the side jambs are to be dadoed. (For the purpose of simplification it is assumed that the side jambs are to be dadoed.)

8. Lay the two side jambs flat on the bench with the edges touching.

9. Square off a mark to represent the bottom end of the side jambs; then measure a distance equal to the height of the door, say 6'8"; then add ½ inch for rug clearance. This point represents the underside of the head jamb. Square a second mark at this point. For very accurate work a pocketknife should be used for the layout.

10. Measure the thickness of the head jamb, and make a mark at

this point to represent the top side of the head jamb. The space between this mark and the mark for the underside of the head jamb should be exactly the same thickness as the jamb stock. This is the dado-cut layout.

11. Mark the length of the head jamb, which will be the width of the door plus twice the depth of the dado into which each end of the headpiece fits.

12. Use a metal miter box, if one is available, and cut off the bottom ends of the side jambs and both ends of the head jamb.

**Note:** If a metal miter box is not available, a wood miter box should be carefully constructed and saw cuts made to receive the saw. The saw cuts should be exactly square, both horizontally and vertically.

13. Gauge the dado cut to the required depth, usually ⅜ inch.

14. Carefully saw the dado cuts to the required depth and remove the wood between the cuts, using a ¾-inch chisel (wider if the dado permits). Finish the cut with the router plane, being sure to work in from both edges of the jamb to prevent splitting the edges.

***Cutting in a Hinge Seat.*** (See Figure 11.) Cutting a recess, or "dap," in a side jamb to receive the leaf of a hinge is much easier if it is done before the jamb is assembled. The tools required are a butt chisel (a short chisel, 1½ inches wide, specially made for cutting in hinges), a marking gauge, pocketknife, router plane, rule, and hammer. The steps involved in the process are the same, irrespective of the size of the hinge. Proceed as follows:

USE KNIFE FOR LAYOUT MARKS

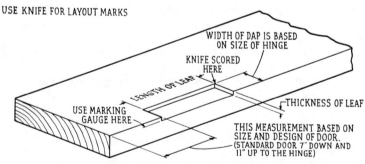

**Figure 11.** Layout for a hinge seat.

1. Determine from the floor plan which way the door is to swing. This is important as it determines which jamb should receive the hinge-seat cut and which edge of the jamb to work from.

2. Measure 7 inches down from the underside of the top dado and 11 inches up from the bottom end of the jamb. These points represent

the standard hinge locations for a standard door. If a third hinge is used (and this is often done for exterior doors), it is centered between the top and bottom hinges. Using a pocketknife, square a mark at these points.

3. Place one hinge leaf below the 7-inch mark and the other above the 11-inch mark; locate and knife-mark the other end of each hinge leaf.

4. Set marking gauge to the width of the dap, usually $1\frac{1}{8}$ inches for a $3\frac{1}{2}$- by $3\frac{1}{2}$-inch butt. This width will vary with the size of the hinge; hence no stated measurement can be given. Gauge a horizontal line from end mark to end mark.

5. Using a knife, deeply score this gauge mark. This will completely sever the wood and make the chiseling work quite simple.

6. Set gauge to hinge-leaf thickness and gauge a line on the edge of the jamb between hinge cross marks.

7. Chisel out the wood between the marks. First cut about $\frac{1}{16}$ inch inside the end marks. Then scarf-cut; hold the chisel at an angle and make several cuts. This results in cutting the wood into small pieces.

8. Carefully remove the wood that has been cut, using a paring cut, in which the back of the chisel is kept perfectly flat and parallel to the face of the jamb. Cutting should follow the depth gauge line.

9. Holding the chisel perfectly plumb, remove the wood at the ends of the recess. The chisel will stay on the knife line since there is no side pressure on the blade of the chisel.

10. Try the hinge leaf in place. If the preceding work has been carefully done, the hinge leaf should be exactly flush with the face of the jamb.

**Note:** Hinges are made with clearance between the hinge leaves when the hinge is closed. This space is the width of the crack between the edge of the door and the face of the jamb. A door will not be hinge-bound if each hinge leaf is set exactly flush with the wood.

The above instructions are applicable to any hinge-seat cutting, irrespective of where the hinge is to be applied.

*Assembling a Doorjamb.* Before assembling the jamb, smooth both edges, planing them slightly "under," that is, out of square, so as to make the casing fit tightly to the jamb; then remove all bruises and marks with a scraper or fine sandpaper. It will save time and effort if gauging for the margin of the casing is also done before the jamb is assembled. Gauge each edge of the jamb from the face side of all pieces for the casing margin, which is normally $\frac{1}{4}$ or $\frac{3}{8}$ inch.

The jamb is now ready to assemble. Using 8d finish nails, nail through one of the side jambs into the end of the head jamb. Repeat for the other jamb. Then nail through the head jamb into the side jamb. The nails must be slanted as shown in Figure 10D. Nail a tie about 6 inches from the bottom of the side jambs. This piece, which is nailed on only temporarily, serves to keep the jamb parallel until it is set. Measurements on the tie are taken from the head jamb.

**Note:** When jambs are mitered at the corners or when a rabbeted jamb is both rabbeted and mitered, the basic instructions for making and assembling the inside doorjamb are the same as given above. Making the miter cuts requires considerable skill since, ordinarily, a miter-box saw cannot be raised high enough for a piece of jamb stock to pass (on edge) under the blade and hence a fine, say an 11-point, finish saw will be required to make the cuts. It would be advisable for a beginner to make a few practice cuts before attempting to make a finished jamb. A miter square is also necessary when making the miter layout.

*Setting an Inside Doorjamb.* (See Figure 12.) The accuracy attained in setting a doorjamb affects several carpentry operations or jobs which follow. If the jamb is set out of plumb or the head is not level, the casing joints will not fit when ordinary methods of square and miter cutting are used. If the side jambs are not set perfectly straight, much difficulty will be encountered in fitting and hanging the door. If the jamb edges are not set perfectly flush with the face of the plaster, the casings will not fit tightly to the edges of the jamb. Hence it is very essential that extreme care be used in setting a doorjamb.

If the layout for a partition-door opening has been done carefully, the doorjamb can be centered in the framed opening. This is necessary when, for instance, the side casings need to be scribed to the wall because a hall is not wide enough to receive two full-width door casings.

Shingles are an excellent material to use for blocking a doorjamb to its correct location, as two of them, used together, can be adjusted to make either a thin or a thick parallel block. Blocks or wedges are needed at the top and bottom ends of both side jambs; on the hinge side, blocks must be placed behind the hinge location (7 inches down and 11 inches up); on the lock side, a block must be placed at the strike-plate location, usually 36 inches from the floor to the center of the plate. Other blocks are used as required to hold the jamb in perfect alignment.

To set a jamb, proceed as follows:

1. Measure the width of the rough or framed opening.
2. Measure the outside width of the doorjamb.
3. Divide the difference between the width of the rough opening and the over-all width of the jamb by two. This gives the thickness of the first set of wedges or blocks.
4. Secure material (shingles) about this thickness, and cut two blocks. The width of the material should be about 4 inches, and the length of the material should be the same as the thickness of the plastered wall.

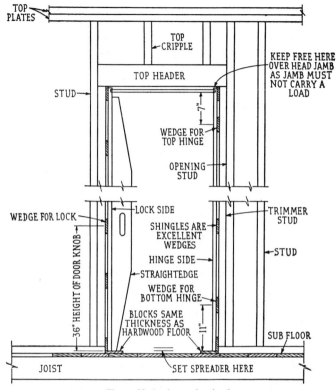

**Figure 12.** Setting a doorjamb.

5. Nail one of the blocks near the top end of one of the door trimmers.

6. Nail the second block at the bottom of the same trimmer. This block should be exactly plumb with the top one. Test it with a straightedge and level. If it is not plumb, either plane off the required amount on the high block or build out the low block (or adjust shingles).

7. Place the jamb in the opening. The bottom end should rest on the finished floor. If the finished floor has not yet been laid, raise the jamb by resting it on strips of exactly the *same thickness as the finished-floor stock*. Test the head jamb for levelness. If necessary, adjust the side jambs and cut off the bottom end of the high side.

8. Nail the side jamb to the blocks on the trimmer. Make sure that the edges of the jamb are flush with the face of the plastered wall. Divide the difference in width, if any, between both wall faces. Use two 8d finish nails per block, slanting them slightly to give additional

holding power. Keep the nails about ¾ inch from each edge of the jamb.

9. Cut a spreader to the exact length of the inside measurement, as measured on the head jamb. The width of the spreader should be about the same as the width of the jamb stock.

10. Place the spreader on the floor against the side jamb first set and then wedge behind the second jamb until it fits tightly and squarely against the spreader. Then nail the jamb to this position.

11. Using a straightedge as illustrated, set jamb blocks or wedges at intermediate points between the top and bottom of the side jamb. Constantly test the jamb as it is wedged and nailed to be sure that it fits perfectly against the straightedge. A "low" point can be easily adjusted by driving the wedge a little tighter. Be sure that the jamb is properly nailed at the hinge locations and the striking-plate location.

**Note:** A doorjamb should *never* be wedged at the top, for the block would transmit a load from the door-opening framed header to the doorjamb, thus causing the door to stick. In other words, a doorjamb must be "free"; it must never carry a load.

12. As a final check, test the jamb with a steel square to be sure that both right angles (head jamb with either side jamb) are perfectly square. If the plumbing and straightedge work has been done accurately, the jamb should be square.

**Note:** If doorjambs are set before the house is plastered, the job of setting will be simplified, provided that the door trimmers were carefully selected at the time the walls were framed. Sometimes these trimmers are only tacked to place when the openings are framed. It is then possible to set the trimmers perfectly plumb; nail one side of the doorjamb (without wedging) directly to this trimmer; then repeat the process for the opposite trimmer stud and doorjamb. Any wedging necessary is done *between* the door trimmer and the full-length stud to which it is framed.

## CUTTING, FITTING, AND NAILING WALL TRIM

*Fitting Window Stool.* Window stool is of two types, casement and double-hung, as shown in Figure 2. The beveled rabbet on the lower side of either pattern is made to conform to a standard window-sill pattern, so that the part of the stool which projects into the room will be exactly level when it is fitted into its place and fastened to the sill. If the sill bevel is not standard, then a standard piece of window stool will not fit properly to the sill.

The standard width of a piece of window stool is 3½ inches. Either type of stool, casement or double-hung (see Figure 2), is designed to project into the frame until the square edge of the rabbet (on the lower face of the stool) fits tightly against the inside edge of the sill. The casement stool, being molded, needs no cutting, other than to make the ends fit against the face of the wall and provide a support for the side casings. The double-hung stool is made to fit *against* the face of the sash and hence must be ripped to attain that result.

It is necessary to hold the stool in a level position while laying it out to make it fit between the jambs. This is done by tack-nailing two pieces of shingle to the sill and letting them project into the room, as shown in Figure 13. Then set the stool rest *in a level position* on these strips and proceed with the marking.

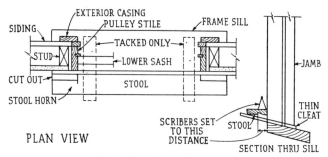

**Figure 13.** Window-stool fitting.

The net length of a window stool is made the width of the window frame, plus twice the width of the casing, plus two "returns," plus casing margins, if any (see Figure 14). A *return,* technically speaking, is that part of the stool which projects beyond the *edge* of the casing and which measures the same as the stool projection beyond the *face* of the casing. The end of the stool is cut the same shape as the front edge of the stool. This is accomplished either by mitering on a return or, preferably, by cutting the stool ends to the same profile as the front edge. The latter method eliminates a very short-grained miter joint on the return, which splits very easily when nailed (see Figure 15, which illustrates an apron return). Stool returns are made in a similar way.

The stool layout is done with a try square and a pair of scribers. Square marks are made to represent the inside face of each jamb. The scribers are used to mark each end of the stool so that it will fit tightly against the face of the plaster (see Figure 13).

**Note:** On a double-hung window it is necessary to fit the sash into the frame *before* fitting the stool (see page 302 for a description of this job). The lower sash is pulled down so that the lower rail rests on the temporary stool sup-

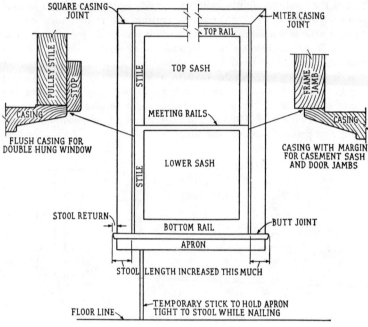

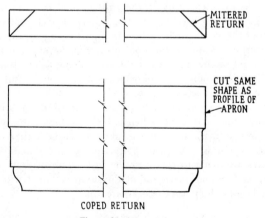

**Figure 14.** Interior-casing joints.

**Figure 15.** Apron return.

port (two shingles). The stool is then scribed and ripped its entire length; the result is a perfect parallel joint between the inside edge of the stool and the inside face of the sash.

In the following summary of the steps to follow when fitting window stool, it is assumed that, if the window frame is double-hung, the sash have been fitted into the frame.

1. Select a piece of stool stock of the right design (either double-hung or casement) and cut to rough length.

2. Tack temporary stool supports to the window sill.

3. Place the stool in a level position, being sure that the same amount of stool projects beyond the edge of each side jamb.

4. Carefully mark the inside face of each side jamb.

5. Measure from the square edge of the stool rabbet to the face of the wall and set the scribers to this amount.

6. Carefully scribe a mark on that part of the face of the stool which is to fit against the plaster. (If a double-hung sash is to be used, mark the entire length of the stool.) Be sure that the scribers are placed exactly at right angles to the face of the wall and kept that way during the marking procedure. If the scribers are permitted to twist, the stool will not fit after cutting.

7. Mark the location of the outside edge of the casing. Add to this the amount that the stool projects beyond the face of the casing. This locates the end of the stool (see Figure 14).

8. Place the stool on a sawhorse and carefully cut on all lines. Use a backsaw for the jamb cuts and a fine-tooth ripsaw for cutting lengthwise.

9. Plane the edge of the stool which fits against the lower sash of the double-hung window. (The casement-sash stool is not planed, as it is molded and the casement sash fits on top of it.)

10. Cut the stool ends so that they have the same profile as the front edge of the stool. This may require a coping saw, or it can be done in a miter box. Mitered returns are usually unnecessary (see Figure 15).

11. Sandpaper the ends to a smooth finish.

12. Remove the shingle strips and place the stool in position. If the job has been carefully done, a perfect fit should result.

13. If the stool is being fitted for a double-hung window, slide the lower sash down to the sill and, if necessary, plane the stool to give proper clearance, so that it will not rub after the sash has been painted.

14. Test the angle between the stool and the wall with a try square to be sure that it is exactly square. If the stool rabbet does not conform exactly to the slope of the sill, it will have to be planed in order to make the part of the stool on which the casing is to rest at right angles to the framing jamb.

15. Nail the stool to place, using 8d finish nails and placing them at least 8 inches o.c. The width of the window frame determines the number of nails to use. Nail about 1 inch from the jamb. (Nail holes may need to be drilled to prevent splitting.)

*Fitting a Window-frame Apron.* The *apron* is the finish piece of trim placed under the stool to cover the inside edge of the frame sill. It also serves as an additional support for the stool and gives a finished effect to the window trim (see Figure 14).

The profile of the apron conforms to the profile of the side casings, as shown in Figure 2, the only difference being that the apron is square-edged to make a tight fit against the stool.

The over-all length of the apron is made the same as the distance from the outside edges of the two opposite side casings (see Figure 14).

The ends of the apron are "returned" by either coping or mitering, as shown in Figure 15. The coped joint is considered the best unless the contour of the molding is such that it makes the job of sawing too complicated. This is rarely the case.

To cut and fit an apron, proceed as follows:

1. Rough-cut a piece of apron stock to fit the window to be trimmed (presumably the stock is already rough-cut, as described on page 266).

2. Determine the exact over-all length of the apron, which is the width of the frame, plus twice the width of the casing stock, plus the casing margins, if any (see Figure 14).

3. Cut the apron to this length.

4. Cut returns by coping or return-mitering.

5. Slightly underbevel the top edge to make a tight joint.

6. Nail the apron to the wall. To hold the apron while nailing, cut a small stick slightly longer than the distance from the underside of the apron to the floor, as shown in Figure 14, and use it to wedge the apron to place; then nail, using two 6d finish nails at each end and spacing them about 10 to 12 inches o.c. Also nail through the stool into the top edge of the apron, using 8d finish nails.

*Casing Doorjambs and Window Frames.* In the following description, it is assumed that the casing stock has been rough-cut to length to make the side and head casings. Doorjambs require four sides and two heads; window frames require two sides and one head. It is advisable to make a careful selection of the best material for the most exposed places; if poor-grained material, which splits easily, must be used, it should be placed in the less conspicuous places—closets and halls, for example.

The steps required to fit and nail casings to either a doorjamb or a window frame are basically the same once the stool and aprons are fitted and nailed on. Doorjambs and casement windows require casing margins; double-hung frames are cased flush, and the joint is then covered with the window stop.

The joints at the top of the doorjamb or window frame, where the side casing is joined to the head casing, may be one of two types, namely, square (butt) or miter, as shown in Figure 14. For all molded trim the miter joint is required. (It should be remembered, as previously explained, that no casing is required when the doorjambs have molded edges or when the plaster is bullnosed.)

FITTING SIDE AND HEAD CASINGS (MITER JOINT). Side casings are kept above the subfloor a distance exactly equal to the thickness of the hardwood floor. If the finish floor is already laid (when pine flooring is used, it is often laid during the framing process and covered with building paper to keep it clean), the casings rest directly on the floor.

Proceed as follows:

1. Make a ¼-inch gauge line on all edges of the doorjamb as a guide for the casing margin (see Figure 14). The gauge line is most easily made at the time the jamb is constructed, unless a margin gauge block is to be used (see Figure 9B).

2. Make a small block the exact thickness of the hardwood flooring. Specifications should be read to be sure of this thickness, for the floor layer must lay the hardwood flooring *under* the door casings.

3. Using a miter box, square one end of a side casing.

4. Lay the flooring block on the subfloor, place the squared end of the casing on the block, and hold the casing tightly against the edge of the jamb.

5. Using a knife, mark the inside edge of the casing at the gauge line on the head casing.

6. Using the miter box, saw a 45-degree-angle cut at the mark, being sure that the mark is the short point of the miter.

7. Using a sharp block plane, very lightly plane the saw cut to a smooth finish. Test the cut for squareness with the face of the casing.

8. Tack-nail the casing to the jamb, keeping the inside edge exactly on the margin gauge line. Drive one nail near each edge about 6 inches up from the bottom, and drive two nails the same distance down from the top end of the casing. Nails will vary in size; for an average 1 by 3 molded casing, use 4d finish nails on the inside edge and 6d finish nails on the outside edge. When nailing the casing permanently, space the nails about 12 inches apart.

**Note:** Experienced carpenters, using a metal miter box, will permanently nail each casing as soon as it is cut. However, the doorjamb (or window frame) may not be perfectly square, and the miter cuts (if made in a wood miter box) may not be exact 45-degree cuts. Tack nailing permits the removal of the casing so that each miter cut can be block-planed to make a perfect joint.

9. Now proceed in exactly the same way with the opposite side casing.

10. Make a miter cut on one end of the head casing.

11. Hold the head casing to place and carefully mark its exact length, which is measured from the long point of the miter cuts of the side casings.

12. Cut the miter.

13. Place the casing in position; check each miter joint, and block-plane where needed.

14. Complete the job of nailing. Nail miter joints through the edges in each direction, using a small finish nail of such size that the wood will not be split.

When a double-hung window frame is being cased, the only difference in the above procedure is that no casing margins are made (see Figure 14). Be sure, if the frame has sash pulleys and weight pockets, *not* to drive nails into the removable weight pocket (see Figure 53, Chapter 7).

**Note:** When sash cord and sash weights are used, the cord should be threaded through the pulley, the weights tied on, and a knot made in the other end of the cord, to prevent it from slipping through the pulley, *before* the window frame is cased. Otherwise this job will require the removal of the weight pockets (see page 232).

FITTING SIDE AND HEAD CASINGS (SQUARE JOINT.) When a head casing is used which is not the same shape as the side casing, a butt joint is required where the two casings meet. Head casings are square and often slightly thicker than the side casings, unless a neck mold is used on the lower edge of the head casing.

**Note:** Today, neck-mold construction is only done on repair work, where the new work must match the old. No information will be given on the use of the neck mold since a beginning carpenter can easily ascertain how to do the job by studying a door or window trimmed in that manner.

To trim a window frame or doorjamb where a square (butt) joint is required, proceed as follows:

1. Cut one end of a piece of side casing square and smooth with a block plane.

2. Place the casing in position; if it is for a window frame, check the fit of the joint where it fits on top of the window stool.

3. Mark the top end of the casing for length. If the casing is for a doorjamb, mark on the margin casing line; if it is for a double-hung window, mark at the inside face of the head jamb.

4. Place the casing in a miter box and cut; then block-plane to a smooth finish.

5. Tack-nail to position.

6. Repeat these operations for the opposite side casing.

7. Square one end of the head casing, and smooth it with the block plane.

8. Mark for length by measuring from the outside edges of the opposite casings. Then add ½ to 1 inch, as it is customary to let the head casing project slightly beyond the side-casing edge.

9. Cut to length on this line and block-plane smooth.

**Note:** A square or flat head casing is never "backed out" by machine as the ends are always exposed. Sometimes it is necessary to back out the casing if the plaster projects beyond the edge of the jamb. This is done with a scrub plane; enough wood is removed to permit the head casing to lie perfectly flat against the wall with the face of the casing aligned with the face of the side casings.

10. Place casing in position and check casing joints.

11. Remove side casings and plane where needed to make a tight joint.

**Note:** The head casing can be nailed first; the "preacher stick" (see Figure 9) is placed over the side casing and against the underside of the head casing. A knife mark or sharp pencil mark is then made on the top edge of the preacher. Cutting slightly "long" on this mark will make a perfect joint, even though the frame may be slightly out of square.

The side casing is placed under the head casing and carefully pushed over the face of the stool (or the hardwood floor block), resulting in a very tight joint.

12. Toenail through the top edge of each end of the head casing with 8d finish nails to hold the butt joint tightly in position. Use 6d finish nails to hold the inside edges of the casing to the jambs, unless the stock is thin, in which case 4d finish nails are large enough.

SCRIBING A CASING. (See Figure 16A.) Sometimes a doorframe will be set so close to the inside angle of a room that it is not possible to use a full-width side casing. When this occurs, the casing must be scribed and cut to fit the surface of the wall against which the casing edge will fit. The casing cannot be ripped to an exact straight line parallel to the inside edge of the casing, for the plastered wall may be slightly uneven, or the jamb, when plumbed properly, may not be exactly parallel to the wall surface.

The scribing job is very simple. The top miter cut is not made until the casing is scribed and cut. Proceed as follows:

1. Select a piece of casing that is perfectly straight. This is important.

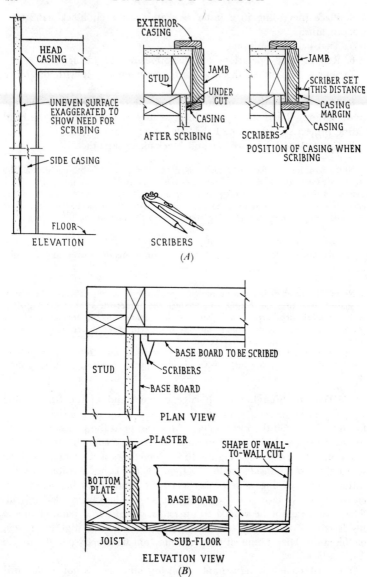

**Figure 16.** Scribing casing and baseboard.

2. Finish-cut the bottom end of the casing to fit the face of the stool, if a window is being trimmed, or to fit the hardwood-floor block, if a doorjamb is being trimmed.

3. Set the casing in position, as shown in Figure 16A, being sure that the inside edge of the casing is parallel with the inside face of the jamb. The distance from jamb face to casing edge is not important, except that the scribers must be set to this measurement.

4. Tack the casing in position to prevent it from moving or falling.

5. Set the scribers to the amount the casing projects beyond the face of the jamb, plus casing margin, if any.

6. Carefully scribe a line the full length of the casing, being sure to keep the scribers in a level position. The metal leg of the scriber will follow the uneven contour of the wall surface; the pencil mark will, correspondingly, be wavy.

**Note:** All wall surfaces are not uneven or out of plumb; however, the casing joint must be properly marked to assure a good joint. The scribing process, if correctly followed, will always result in a perfect casing-edge fit.

7. Rip the casing on the line, undercutting slightly so that the casing edge, when ripped, will be sharp. Undercutting allows for any "out-of-squareness" in the wall angle.

8. Mark the top end of the casing where the miter or butt joint is to be made and make the finish cut.

9. Block-plane this finish cut.

10. Nail the casing into place.

*Locating Wall Studs by Sounding.* In order to ensure solid nailing for wide baseboard, hook strips, or picture molding, it is necessary to locate the wall studs. This is done by tapping the wall lightly with a hammer. A solid sound will be heard at the stud locations. The wall surface between the studs will produce a hollow sound.

By placing the fingers of the left hand near the surface that is being tapped, one can feel vibrations when there is no stud. This "touch" system is very helpful, particularly if plaster lath has been used, for it is otherwise very difficult to locate a stud after a wall has been plastered.

Once a stud has been found, other studs are located by measuring 16 inches in either direction from this stud, as studs for plastered walls are spaced 16 inches o.c. to conform to the length of plaster lath, which is 32 inches, or of wood lath, which is 48 inches.

For a narrow base or for picture molding that is to be placed at the ceiling line, no sounding is necessary, as the top and bottom wall plates are continuous; hence the base or molding can be nailed anywhere.

*Fitting and Nailing Baseboard.* \* As explained previously, baseboard is sometimes fitted and nailed on before the plastering is done. The job of fitting is quite simple as inside- and outside-angle joints can be made perfectly tight in a miter box.

When a molded baseboard is used and nailed on after the plastering is done, it becomes necessary to scribe all inside angles in a manner somewhat similar to the method followed when scribing a casing.

Outside angles require a miter joint, which, on narrow base, can be cut in a miter box.

In most rooms there will be one or more walls in which there are no door openings, requiring that the baseboard be fitted between two opposite walls. If the wall is over 20 feet in length, two pieces of base are required, and a butt joint is made on a stud to provide proper nailing. (Inside-finish lumber rarely exceeds 20-foot lengths.)

When one piece of baseboard is fitted between opposite walls, two rods are used to measure the exact length of the base.

When there are door openings in a wall, the base is fitted by working from the inside angle. This method gives a free end to the base, which can then be easily measured for exact length at a door casing, after a scribed joint is cut for the inside angle of the room.

FITTING BASEBOARD BETWEEN TWO WALLS. (See Figure 17.) To fit a piece of base between two walls, proceed as follows:

1. Secure two pieces of 1 by 1 material, each of which is a little shorter than the between-wall measurement.

2. Select a piece of base long enough to go from wall to wall. Cut one end slightly out of square, as shown in Figure 16, to make the top edge fit tightly to the plaster.

3. Hold the two rods tightly together, with one end of each rod against the wall. The measurement should be taken at the top line of the baseboard. While holding the rods in this position, make "witness" marks, as shown in Figure 17. If the combined rod is rather long, tack-nail to prevent it from slipping.

4. Place the rod on the piece of base; keep one end even with the top edge of the out-of-square cut, and mark the base at the other end of the rod.

5. Cut the base on this mark, again making an out-of-square cut (known as an undercut).

6. Place the base in position by holding the center away from the

---

\* Base shoe is fitted the same as picture molding (see page 292), except that the carpenter works to the door openings to provide a free end for the last piece of shoe. Coped joints are used for inside angles; coped returns are used where the molding fits against a door casing.

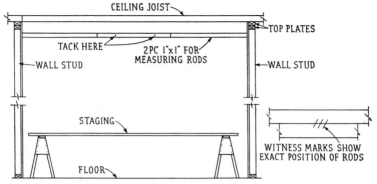

**Figure 17.** Rod usage when fitting picture molding (baseboard is similar).

wall and springing the ends into place first. This will prevent scratching or marring the surface of the finished plaster wall.

7. Nail to place at each stud, using 6d finish nails for thin base and 8d finish nails for ¾-inch base. Wide baseboard requires two nails per stud, one near the top edge and the other near the bottom edge of the baseboard. All nails should be set to receive putty.

SCRIBING BASEBOARD. (See Figure 16*B*). To scribe one piece of baseboard to another, proceed as follows:

1. Select a piece of baseboard long enough to go from the inside angle of the room to the door opening. (Presumably the base has already been rough-cut to length.)

2. Place the base in position by resting it on two hardwood-flooring blocks to keep it parallel to the subfloor and butt one end against the base already nailed on.

3. Set scribers as illustrated and carefully scribe a mark on the end of the base.

4. Cut the base to this line, which is actually the exact profile of the baseboard. On curved portions of the line a coping saw will be needed. Be sure to undercut, thus giving the cut a sharp edge, which will ensure a perfect fit.

5. Place the base in position, check the joint for fit, and block-plane or cut where necessary; then, using the preacher, mark the base where it fits against the casing (see Figure 9*C*).

6. Saw to this line, giving it a very slight undercut.

7. Block-plane this joint lightly to make it perfectly smooth.

8. Place the base against the casing and carefully push the baseboard into position. If the piece is too long, a plaster crack will be made at

the wall angle; if it is too short, the joint will not be tight; hence all measurements must be carefully taken.

Note: It is recommended that a beginner work first in a clothes closet, where the baseboard joints are not so readily seen. A little experience will soon give him the skill necessary to do a first-class job of finish work.

MITERING BASEBOARD FOR AN OUTSIDE ANGLE. Before marking the baseboard for the miter cut, test the angle for squareness; as an out-of-square corner requires adjusting the length of the base to ensure that the baseboard will be fitted at a 90-degree angle. In other words a "short" wall, as shown by the steel-square test, requires that the baseboard be cut longer to allow for this shortness.

Note: All outside angles and plastered archways in a plastered wall should be carefully protected with metal corner beads. The bead is nailed on prior to plastering and acts as a guide for the plasterer as well as protection for the corner; hence it must be carefully nailed on. A straightedge should be used to test its straightness sideways and along the front edge.

To make a miter cut, proceed as follows:

1. Place the base in position. It is assumed that one end of the base has been fitted and that the piece is ready to cut to exact length.

2. Using a sharp and reasonably hard pencil, mark the location of the wall angle on the back side of the baseboard.

3. Using a miter square, transfer the mark to the face of the base by making a miter-cut mark on the top edge of the base.

4. Carry this line across the face of the base.

5. If possible, cut the miter in a miter box; otherwise lay the base on a pair of horses and saw the miter, being sure to work from the face edge.

6. Block-plane the joint to a smooth cut.

7. Nail the base to place.

8. Repeat this procedure for the other pieces of base which are to fit around the corner. On a plastered archway three pieces of base will be required.

9. Using small finish nails, carefully nail the miter in both directions to hold it to place and prevent it from opening up.

Note: If the wall angle is not a perfect 90 degrees, it may be necessary to fill in the space behind the top edge of the baseboard. While the need for this indicates poor craftsmanship, either in wall layout or in plastering, it is far better to fill in the crack so that the base will show a perfect angle than to attempt to make the baseboard follow the shape of the wall.

*Fitting and Nailing Hook Strip.* Hook strip is a flat piece of ⅝- by 4-inch finish material with two rounded corners (see Figure 3), into which clothes hooks are screwed. (See Chapter 9, page 349, for information on the spacing of clothes hooks.) The trim is nailed horizontally on the walls of a clothes closet, with the top edge of the hook strip approximately 5′6″ from the floor.

The construction procedures to follow when fitting hook strip are identical with those described for fitting baseboard. Inside angles are scribed; outside angles are mitered, and the preacher is used to mark the square joint required where the hook strip butts to a door casing.

Care must be used to make sure that the hook strip is level. Cut a small rod the correct height and rest the hook strip on it while nailing. Even though a supporting rod is used, it is best to check the hook strip with the level since it is nailed on at eye height. Two 8d finish nails must be driven into each stud.

Closet shelving is often nailed on top of the hook strip; clothes-pole rosettes are nailed to the face of the hook strip.

***Fitting Clothes Poles.*** Clothes poles are usually made of wood and are 1½ inches in diameter, though 1¼ inches is large enough for short poles. They are fastened to place by one of two methods: (1) by boring a hole through the hook strips that are to support each end of the pole and then carefully pushing the hook strips to their intended place with the pole already in the holes; or (2) by using a rosette, as illustrated in Figure 3. (Plastic rosettes are also available.) The latter method is the simpler. One rosette is nailed on the face of the hook strip, and the pole is inserted into the second rosette prior to nailing it to place.

Clothes poles must be placed at least 10 inches out from the wall to provide clearance for the clothes hangers. The position of the pole is determined by the shape of the closet; it is usually an advantage to make the pole as long as possible. Many clothes closets have two poles, one at each end, running across the narrow dimension of the closet.

***Fitting Closet Shelving.*** Some closet shelves are held up by the hook strip, which acts as a support for each end of each shelf. Many homes, however, especially those with small closets and limited storage space, utilize the space above the hook strip for additional shelves. These shelves are usually made from 1 by 12 shelving lumber and are supported by 1 by 2 cleats. An important construction point needs emphasis here: a shelf cleat is only as strong as its nailing into the wall. If no stud nailing is available at the outer end of a shelf cleat, vertical supports must be fitted between each cleat. If the shelving requirements of each clothes closet are known *at the time the walls are framed,* additional backing should be framed in to provide the necessary end-nailing support for the shelf cleats.

The location of the shelf cleats should be carefully planned on the basis of the closet storage possibilities.

Shelf cleats are cut the same length as the width measurement of the shelf. Cutting the outside end on a 45-degree angle eliminates the blunt-end appearance of a cleat.

All cleats should be hand-smoothed and sandpapered before they are nailed to the walls. The lower exposed edge of each cleat should also be slightly beveled to eliminate the sharp edge.

It is necessary to use 8d finish nails to fasten shelf cleats to a plastered wall, for the nails must be long enough to enter the studs and hold the cleats firmly to place. One nail at each end is sufficient.

The procedures required to fit a shelf between two closet walls are as follows:

1. Measure the approximate distance between the two walls and rough-cut a piece of shelving at least 1 inch longer than this measurement. Rough-cut refers to length; a finish saw is used to make the cut.

2. Check the wall angle with a steel square and cut one end of the shelf accordingly. In all probability the plastered angle will be "full," requiring that the inside end of the end cut be cut "short," as shown by the steel square.

3. Using an extension rule (or two sticks), take a careful measurement between the walls at the location of the front edge of the shelf.

4. Mark the piece of shelving and cut it to this length, again making the cut out-of-square, as shown by the steel-square check.

5. Plane the front edge of the shelf smooth and straight.

6. Using the smoothing plane (or any small plane), slightly bevel the two corners of the front edge; then round off with sandpaper so that no sharp corners are left. These rounded edges are appreciated by the home-owner, who will make frequent use of the closets.

7. Nail the shelf to place, if necessary, with 6d finish nails. Quite frequently closet shelves are left loose so that they can be removed when the closet walls are painted or papered.

**Cutting and Fitting Picture Molding.** While many modern houses do not have picture molding, the skilled carpenter must know how to cut and fit this molding when the occasion arises. Basically the procedures are applicable to any molding which is continuous around a room.

Sometimes a picture molding is placed directly in the wall-and-ceiling angle, in which case no guideline is needed to keep the molding straight as it is nailed on. If the molding is placed down from the ceiling to permit picture hooks to be placed on its top edge, a chalk line should be snapped on the wall to simplify the job of keeping the lower edge of the molding straight while it is being nailed on.

Picture molding is nailed at intervals of about 16 inches; if it is placed close to the ceiling, it can be nailed anywhere since the top plates in the framed wall provide adequate nailing. If the molding is placed several inches down from the ceiling, it is necessary to locate the studs (by sounding) for proper nailing. The size of nails used depends on the thickness and shape of the molding; ordinarily 6d finish nails are long enough to give good holding power in the studding.

All inside-angle joints are coped, not mitered. A properly coped molding will never open up, for increased pressure placed on the ends of the molding makes a tighter joint. Excessive pressure will, however, cause corner cracks in the plaster.

Exterior corners are always mitered, and a tight joint is made by nailing through the miters in each direction.

Making a coped joint on the end of a piece of molding necessitates making a back-miter cut (see Figure 18*A*). To make this cut, place the molding in the miter box upside down; that is, place the top edge of the molding so that it rests on the bottom of the miter box. While the molding is in this position, make a miter cut on one end of the molding, *proceeding exactly as you would if the finish joint were to be mitered.* Then, using a coping saw, cope the end of the molding by sawing (making an undercut) on the profile edge of the miter cut. This finished cut will be the exact opposite of the molding profile (see Figure 18*B*).

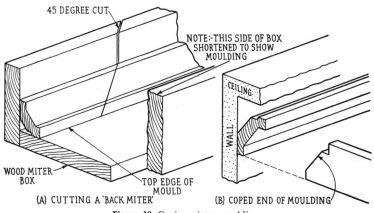

**Figure 18.** Coping picture molding.

In a room with four walls the first piece of molding is cut square on both ends. The second and third pieces are coped at one end only; the other ends are cut square. The fourth and last piece must be coped at both ends and sprung into place. It is advisable to leave the longest molding until last, as it will have more "spring" and hence be easier to place in position on the wall.

Figure 17 illustrates how to determine the length of a piece of molding which is to be fitted between two walls. Two rods of suitable size (not too large) are placed side by side and extended until one end of each rod touches the face of the wall. While they are in this position, carefully make "witness marks," as illustrated, to show the position of the rods while they are lapped. If the rods slip or move during the measuring process, they can be quickly placed together again exactly in their original position.

As soon as the first piece is measured and square-cut, nail it onto the wall (using a chalk line, if necessary, as indicated above).

When measuring the second and third pieces of molding, be sure that the end of the rod is placed against the lowest member of the molding; when measuring the last piece, see that both ends of the rod are placed against the molding, *not* against the plastered face of the wall.

## FITTING DOORS AND CASEMENT SASH

The procedures to follow when fitting either an inside door or a casement sash are basically the same. (Double-hung windows, since they require an entirely different method, will be discussed separately.)

*Making a Door Jack.* If a carpenter has a number of doors or sash to fit, he can save time and do the job much more easily if he first makes a door jack. Two types are commonly used. A very durable door jack is illustrated in Figure 19*A*. The dimensions given on the drawing should be adjusted to suit the dimensions of the door or sash to be fitted. The vertical member should be well braced; the horizontal 1-inch piece with the V cut should be well nailed, with the nails set, for this piece serves as an excellent rest for the jointer plane. This type of jack is worth storing as it can be used on many jobs.

A simplified door jack is shown in Figure 19*B*. In this type two pieces of 2 by 6 are nailed to a wood lath, and the space between the inside ends of the stock is made a trifle wider than the thickness of the door. A 1⅜-inch door would require a 1½-inch space. The weight of the door on the bottom thin member causes it to sag and tip the top edges of the 2 by 6 so that they pinch the door tightly. The 1 by 4 foot piece is merely for the purpose of providing space below the open slot. The "feet" are nailed temporarily to the floor to prevent the jack from moving while the door is being planed.

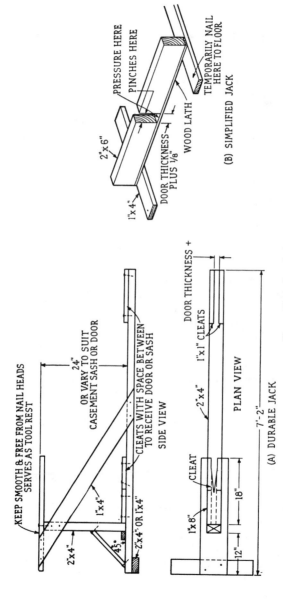

PRESSURE HERE

PINCHES HERE

TEMPORARILY NAIL HERE TO FLOOR

2"x 6"

DOOR THICKNESS PLUS ⅛"

WOOD LATH

1"x 4"

(B) SIMPLIFIED JACK

KEEP SMOOTH & FREE FROM NAIL HEADS SERVES AS TOOL REST

24" OR VARY TO SUIT CASEMENT SASH OR DOOR

CLEATS WITH SPACE BETWEEN TO RECEIVE DOOR OR SASH

SIDE VIEW

1"x 4"

2"x 4"

45°

2"x 4" OR 1"x 4"

DOOR THICKNESS +

1"x 1" CLEATS

2"x 4"

CLEAT

1"x 8"

18"

12"

PLAN VIEW

7'-2"

(A) DURABLE JACK

**Figure 19.** A door jack.

*Fitting an Inside Door.* If a doorjamb is properly set, it should be parallel, square, and straight. The head jamb should be tested with the steel square, the side jambs with a straightedge, and the width measurements, both top and bottom, with a rule. After the doorjamb is cased, it is too late to make any changes in the jamb setting, and the door must then be fitted to the jamb as set.

When a door is hung, the cracks between all door edges and the jamb should be the same size. This crack is made larger for a door with a paint finish than for one with a stain-and-varnish finish.

The lock-edge door stile must be beveled lightly to prevent it from striking the doorjamb as the door is closed. A narrow door will require more bevel.

The bottom edge of the door must have at least ½-inch clearance for rugs. A properly made jamb will be ½ inch longer than the door to provide for this clearance and eliminate the necessity of sawing off the bottom of the door.

A door should swing freely after it is hinged. If the door is "hinge-bound," (1) the hinges have been gained in improperly—too deep or not flush—or (2) the doorstops on the hinge side of the door have been set too tight against the door.

If the jambs have not already been cut for the hinges, the plans should be carefully checked to determine the correct swing of the door. The location of electric-light switches should also be checked to be sure that no light switch is behind a door when the door is opened.

To hang an inside door, proceed as follows:

1. Drive two finish nails into each jamb, one 6 inches from the top and the other 12 inches from the bottom. They must be set in from the edge of the doorjamb an amount equal to the thickness of the door. (These nails are temporary and should be driven in only ½ inch, then removed when the doorstop is nailed into place.)

2. Place the door on the sawhorse, and saw the lugs off the top and bottom stiles. (Use a fine saw and be careful not to tear the ends of the stiles.)

3. With the crowning face of the door placed out, try the door in the jamb opening and note the amount that has to be planed off.

4. Place the door in the door jack, with the hinge edge up, and plane this edge straight and square, using an 18- or 22-inch jointer plane.

5. Try the door in the jamb opening again, placing it against the hanging stile of the jamb to see how it fits. Plane until it fits the jamb, being sure not to make the door too small.

6. Measure the width of the jamb at the top and bottom, transfer these sizes to the door by measuring from the hinge stile, and mark the sizes on the lock stile.

7. Draw a line from these marks and plane to this line.

8. Try the door in the opening to be sure the side joints are parallel.

9. With the door tight against the hinge jamb, scribe the top of the door to fit the top or head of the jamb.

10. Plane or saw this amount off the top rail of the door. Be sure to plane door stiles from the outside edge to avoid splitting the stile.

11. Try the door in opening and plane where necessary. Leave a $\frac{1}{16}$-inch clearance on sides and top. Paint jobs will require a little more clearance.

12. Bevel lock stile slightly.

13. Wedge the door in the opening, keeping the hinge stile against the jamb, and the top of the door $\frac{1}{16}$ inch below the head jamb. Two 4d finish nails placed in the crack at the top end of the door will give the proper clearance.

14. Measure down 7 inches from the top of the door and 11 inches up from the bottom; using a sharp knife, mark these points both on the door and the jamb. (The 7-inch mark represents the top of the upper hinge, and the 11-inch mark the bottom of the lower hinge. When laying off for the hinges, be sure to keep on the correct side of the marks.) Replace the door into the door jack.

15. Lay out the dap or recess for the hinge leaf, as outlined on page 274 (see Figure 11). On a $1\frac{3}{8}$-inch door the hinge seat is cut 1 inch wide in the door stile and jamb.

16. Gain in for the hinge seats on the edge of the door and the face of the jamb.

17. Screw each hinge leaf on the door and jamb so that the hinge pins are pushed in from the top of the hinge. Use beeswax or soap on the screw threads to make them screw in more easily.

18. Set the door in place and push in the pins. Test the door to see that it swings freely.

19. When the door swings freely and is in good working order, sandpaper all the edges slightly to remove all sharp edges.

DOORSTOPS. Doorstops, one headpiece and two sidepieces, are usually $\frac{3}{8}$ by $1\frac{1}{2}$ inch in size. The headpiece is cut in first, with both ends made square to fit tightly against the face of the doorjamb.

The side stops are coped to the head stop. The bottom end of the stop is made flush with the bottom end of the jamb to permit the hardwood floor to pass under. Measure the length from a piece of hardwood flooring placed under the jamb.

The stop should be held to position by 4d finish nails spaced 8 to 10 inches o.c. It should *not* be nailed until after the door lock is fitted; the exact position of each piece can then be accurately determined.

The stop on the lock side is nailed tightly against the door. The stop on the hinge jamb must be kept at least $\frac{1}{16}$ inch away from the face of the door to permit the door to swing freely after it is painted.

HANGING A DOUBLE-ACTING DOOR. Steps 1 through 10 above can be applied in hanging a double-acting door. After the door fits *into* the frame, the front edge of the door is rounded to permit free movement through the jamb. The double-acting door will require at least $\frac{1}{8}$-inch clearance on both edges.

No instructions can be given regarding the application of the double-acting-door hinge as there are many different types and designs. In residential construction the hinge is always screwed to the bottom edge of the door, and a flat hinge plate is screwed to the hardwood floor. (The hinge is raised from the floor with a piece of wood the exact thickness of the hardwood flooring so that the door can be hung before the hardwood flooring is laid.)

Note: There is one type of double-acting hinge which is screwed to the side edge (stile) of the door and not to the bottom edge. This type of hinge is often used in public buildings because of its greater durability. When it is used, the doorjamb must be made at least 1 inch wider to allow room for a hinge strip, in which the hinge is recessed.

The double-acting hinge should always be purchased and delivered to the job before the double-acting door is fitted, and the instructions which are printed and enclosed with every hinge should be carefully read and followed. The shape of the door edges is sometimes based on the shape of the exposed parts of the hinge.

*Fitting Casement Sash (Swing-in Type).* Fitting a casement sash that swings into the room is similar to fitting an inside door, except that the bottom edge of the sash must be accurately fitted to the window stool; this means that all four edges of a sash must be fitted to the window frame.

The bottom edge of the sash is rabbeted to fit over the window stool; the rabbet is usually machined in the mill at the time the sash is made. Consequently, vertical measurements are taken from the face of the stool to the underside of the jamb and transferred to the sash; then the top edge of the sash is planed off to suit these measurements. Width measurements are taken from jamb face to jamb face.

Special hinges, called parliament butts, are used to give the necessary clearance for window-shade and window-curtain brackets when the sash is opened to a 180-degree angle. The front, or opening, edge of a casement sash must be beveled considerably to conform to the swing of this hinge.

The procedure for cutting in the hinge seats is the same as already described.

In summary, the steps required to fit and hang a casement sash which swings into the room are as follows:

1. Measure the over-all dimensions of the sash (both width and length) to determine how much must be planed off the sash to make it fit into the window frame.

2. Check the rabbet on the bottom edge of the sash to be sure it conforms to the lip on the casement stool (see Figure 2).

3. Measure the vertical dimensions of the frame and mark them on the sash.

4. Plane off the top edge of the sash according to these measurements.

5. Determine which edge of the sash is to be hinged, and plane this edge first.

5. Bevel-plane the front edge of the sash; the sash should occasionally be tried in the frame to check the places that need planing and also to determine when sufficient stock has been removed. A paint job will require a larger crack than will a stain-and-varnish job.

7. Locate the hinges 3 to 4 inches down from the jamb and 5 to 6 inches up from the stool.

8. Cut in the hinge seats and screw on the hinge leaves. The width of the hinge seat varies with the type of hinge, but ¾ inch is standard.

9. Place the sash in the frame, and push the hinge pins into place.

10. Test the swing of the sash to be sure that it is not hinge-bound and that it clears the window frame on all edges. Plane where necessary. The beveled front edge may need more bevel—the narrower the sash, the greater the bevel.

11. Plane off all corners lightly and sandpaper smooth.

*Fitting Casement Sash (Swing-out Type).* The same general rules are followed in fitting a swing-out sash as in fitting a sash that swings in. For the swing-out type, however, the lower rail must be beveled to fit the bevel of the sill. A T bevel is set to fit the pitch of the sill, and the bevel is transferred to the bottom end of each sash stile. Care must be taken to see the bevel is made so that the glass putty is outside.

The swing-out sash is more difficult to fit, particularly on a two-story house; wherever possible, the sash should be fitted *before* the window frame is set. It may be necessary to work from a ladder or, on a siding house, from a scaffold.

Hinges for a swing-out sash are often sheradized to make them impervious to rust.

To obtain a full bearing, the hinge leaves are set in for the full width of the leaf. A casement-sash adjuster is used to hold the sash from swinging in the wind; some types of adjusters are fitted through the inside window-frame apron and operated with a crank (see page 342).

Sash which have been fitted and hung before the job is plastered must be taken off; this is done by merely pulling the hinge pins.

The sash should be primed as soon as hung; the top edge of the top rail and the top ends of the stiles should be thoroughly painted to prevent moisture from entering the wood.

*Fitting and Hanging Interior Sliding Doors.* Until a few years ago it was necessary, when framing for a sliding door, to erect *two* parallel stud walls with sufficient space between them to receive the sliding-door track and the sliding door. This meant that the over-all thickness of the two walls had to be in excess of 10 inches. Modern developments, however, have overcome the need for dual walls, and it is now possible to prepare a 4-inch stud wall to receive a sliding door. One such development is known as the Glide-Master. The following information and accompanying illustrations describe the framing and installation procedures for the Glide-Master,* which are discussed here as one operation.

Figure 20 illustrates the Glide-Master Door Frame, which is bought as a complete unit ready to install in a framed wall. For a 6'8" door the bottom edge of the opening header is framed 7'0½" from the subfloor. The width of the framed opening is double the width of the door, plus 2 inches. To illustrate, a 2'6" by 6'8" door would require a framed opening 5'2" by 7'0½".

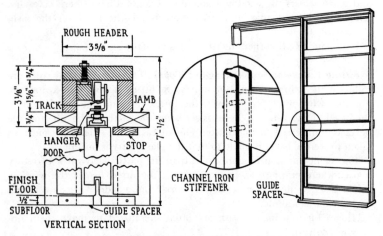

**Figure 20.** Glide-Master sliding-door frame.

* Information and illustrations pertaining to the Glide-Master Door Frame and hardware have been provided through the courtesy of Arthur Cox and Sons, Pasadena, California, manufacturers of sliding-door and wardrobe hardware.

The Glide-Master Door Frame is placed in the framed opening, which has sufficient width and height clearance for the frame to be set perfectly plumb and level. This is very important; on a sliding-door track set out of level the door would not stay put.

When the doorframe is level and plumb, it is toenailed to the subfloor through the bottom plate of the frame and toenailed, at the top, to the top header.

To clarify the installation procedure, the description of the hardware for this door, shown in Figure 21, is included in the following instructions on how to hang the door:

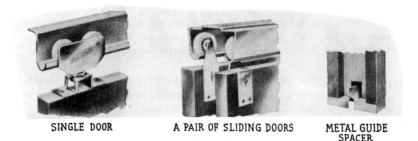

SINGLE DOOR          A PAIR OF SLIDING DOORS          METAL GUIDE
                                                         SPACER

**Figure 21.** Glide-Master sliding-door hardware.

1. Cut the door to 6'7¾" to allow ½-inch clearance between the finish floor and the door (based on a ½-inch finish floor). Groove the door bottom, ¼ inch wide and ¾ inch deep, to allow clearance for the Guide Spacer.

2. Screw doorplates in the center of the top rail of the door, 3½ inches in from each edge. The doorplate slots face each other (see Figure 22).

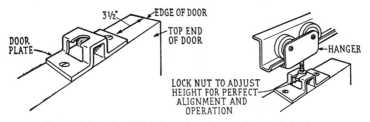

**Figure 22.** Locating Glide-Master doorplate on the top of the door.

3. Attach bolt and lock nut to hanger and install in track. Screw bolt up to track; then back off two full turns for clearance.

4. "Position" the door in the opening by inserting the door groove in the Guide Spacer and sliding the door in until the doorplate aligns with the header cutout (a band-sawed curve to provide room to insert the door hardware). Slide the hanger bolts into the doorplate, and test the door for alignment. One full turn of the hanger bolt adjusts the door $\frac{1}{4}$ inch. With the door operating and accurately aligned, tighten the lock nut to the doorplate (see Figure 22).

*Fitting a Double-hung Window.* A double-hung window consists of two sash, upper and lower, which slide vertically and are balanced by using either sash weights or sash balances (see Figures 37 and 53, Chapter 7).

The top sash is fitted first; the lower sash is then fitted so that the meeting rails of the two sash (see Figure 15) are exactly flush and the lower rail of the bottom sash fits snugly to the pitch of the window-frame sill.

Each sash must be fitted rather loosely to permit it to slide freely. All corners of the outside edges of the sash, except the meeting rails, should be beveled slightly to eliminate the sharpness.

The parting bead must be removed to permit the top sash to be fitted into the space between the parting bead and the blind stop (see Figure 32, Chapter 7).

A small piece of each end of each meeting rail must be removed to provide space for the parting bead (see Figure 23). After the sash is fitted, measure $\frac{1}{2}$ inch in from the edge of the sash and, using a fine backsaw, cut the beveled edge of the meeting rail, making a cut $\frac{1}{8}$ inch in depth. Using a wide chisel, remove the part of the rail which protrudes beyond the face of the sash stile (see plan view, Figure 23).

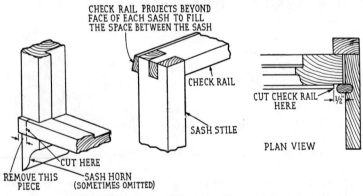

CHECK RAIL PROJECTS BEYOND FACE OF EACH SASH TO FILL THE SPACE BETWEEN THE SASH

CHECK RAIL

SASH STILE

CUT HERE

REMOVE THIS PIECE

SASH HORN (SOMETIMES OMITTED)

CUT CHECK RAIL HERE   $\frac{1}{2}$"

PLAN VIEW

**Figure 23.** Check-rail cutting to allow for parting bead.

To fit the top sash, proceed as follows:

1. Remove the parting bead from the frame.

2. Try the sash for size, and plane the edges until there is sufficient side clearance, at least $\frac{3}{16}$ inch less than the width of the frame.

3. Place the sash into the frame and test the fit of the top rail against the head jamb. Plane accordingly.

4. Slightly bevel all stile and top-rail edges.

5. Replace the parting bead, first the headpiece and then the side-pieces. With the top sash resting on the sill to permit the side parting beads to be sprung into place, slide the lower end of the bead into the space made by cutting a notch in the beveled edge of the meeting rail. A correctly milled parting bead does not require nailing.

6. Push the sash to the top of the frame and hold it there by driving two small nails under the lower edge of the sash horn. This completes the fitting of the top sash.

To fit the lower sash, proceed as follows:

1. Try the sash for size, and plane the edges until there is sufficient side clearance.

2. Place the sash into the frame and hold it in place by driving a small nail into each frame jamb. This prevents the sash from falling out while the bottom rail is being marked.

3. When a lower sash is properly fitted, the top edge of the meeting rail must be exactly flush with the meeting rail of the top sash. The lower-sash meeting rail will be above the upper-sash meeting rail until the lower rail is fitted to the sill. This vertical distance must be carefully measured and scribers set to this measurement.

4. Carefully scribe the outside of the lower rail of the sash. (The inside face of the rail is behind the window stool and cannot be marked unless the sash are fitted before casing the frame.)

5. Set a T bevel to the exact pitch of the window sill and mark on the outside edge of each sash stile. The long edge of the bevel must coincide with the scribed mark on the outside face of the sash.

6. Saw off the stiles and rail to this pitch; cut below the line a little to allow for planing.

7. Plane off the rail to a smooth surface; try the sash in the frame and continue planing until the meeting rails are exactly flush. (Flush rails are the test of a correct job of fitting a double-hung window.) Then bevel the corners slightly.

**Note:** Sash-lock hardware is designed to lock the sash by automatically squeezing the meeting rails tightly together; it also makes the meeting rails flush. A "short" fit on the lower sash is an obvious piece of poor carpentry, for it causes the lower rail to be lifted off the face of the sill when the sash are locked together.

Step 7 completes the fitting of the upper and lower sash in readiness for fastening the sash-balance hardware.

FITTING SASH BALANCES. Since there are several different types of sash balances and since each sash balance is packed with full installation instructions, the only requirement that need be noted here is to follow the printed instructions.

FASTENING SASH WEIGHTS. A pair of fairly accurate spring-balance scales are needed to determine the size of weights that should be used for a window of a specified size. Carefully hook the sash on the scale and read the weight. Then divide this figure in half to determine the size of weights required. For instance, a sash weighing 8 pounds will require two 4-pound weights (see Figure 37, page 214).

Note: Sash vary in weight on the basis of the thickness of glass used. Obviously 16-ounce glass is lighter than 21-ounce glass (weight per square foot). Occasionally the kind of wood used may affect the weight of the sash. When sash are of odd weights, ½-pound "washers" can be bought to bring the sash weights into perfect balance with the sash.

Four sash cords, cut the full length of the window frame, are required per frame. To simplify the hanging procedure, sash cords should be threaded in the pulley and the weights tied on *before* the side casings are nailed on; otherwise it will be necessary to remove the weight pockets to tie on the weights. (Trouble lies ahead if there is no weight pocket!)

In summary, to hang a double-hung window on sash cords, proceed as follows:

1. Weigh the sash and order the weights.

2. Cut four cords, thread them through the pulley, and tie on the weights. Tie a knot on the other end of the cord to prevent it from slipping through the pulley.

3. Fit the sash as outlined above.

4. Prepare the outside edge of each sash stile for the cord knot. A 1-inch-diameter hole has already been made by the sash manufacturer, as well as a ½-inch half-round groove for the cord to fit in. The groove and the hole are connected by sawing a dovetail slot, ¼ inch wide at the top and ⅜ inch wide at the bottom. The cord will squeeze through this tight slot and will not come out after the sash is hung.

Note: To be on the safe side, drive a small flat-headed nail through the knot.

5. Fit the cord into the top sash, place the sash into the frame, and replace the parting bead. Test for balance, and add washers if necessary.

6. Repeat this operation for the lower sash.

7. Fit and nail in the window stop. The top, or horizontal, stop is nailed in first; both ends are cut square to fit against the face of the pulley stile. Sometimes this stop is omitted.

8. Cut in the side stops, making a cope joint at the top and a square joint at the bottom to fit to the window stool.

9. Nail in the stop, being sure not to crowd it against the lower sash; at least $\frac{1}{16}$-inch clearance should be allowed for the paint or varnish. Use 4d finish nails, spaced approximately 8 inches o.c.

**Note:** On costly jobs or public buildings, side stops are often fastened in with small washers and screws. The screw holes are slotted to permit a side movement of the stop. To loosen a stop, it is necessary only to loosen a screw, move the stop slightly, and then tighten the screw.

## CABINETWORK

The term *cabinetwork,* when used in connection with house construction, applies to a variety of built-in features required to make a house convenient and livable. Figure 7 illustrates some of the typical cabinets found in an average residence. Some of them are not exactly cabinets in the true sense of the word—for example, a built-in ironing board, or a linen closet constructed in a plastered recess so that the shelves rest on cleats and no side jambs are required.

From the carpenter's viewpoint, cabinetwork covers all the interior-finish construction of the house except trim work and door, sash, and window hanging. For mass-housing projects, cabinetwork is usually constructed in a mill as a stock item, and the framing layout must be made to fit the stock-item measurements. On a single-residence job, the mill will send a layout man to take the "job measurements," and the cabinets are, in a sense, tailor-made.

A third method is to build the cabinets on the job; finish lumber is ordered in long lengths and varying widths (1 by 8, 1 by 10, 1 by 12, etc.). The various parts of each cabinet are prepared by cutting, jointing, gluing, cleaning, and dadoing; then the cabinet is either assembled from the various parts and fitted to its place or constructed piece by piece in the required location. The fronts of cabinets and the drawer stock are ordered in long lengths as milled items and cut to make specific parts of a cabinet. The information below pertains to this on-the-job construction method.

*Equipment and Working Area.* A finished cabinet reflects the skill of the carpenter who did the job. Proper equipment is a necessity: a smooth-top workbench and smooth-top sawhorses; a variety of finish tools that have sharp cutting edges and are in good working order; cold glue is a necessity for gluing flat stock edge to edge, to fasten a cabinet front to a cabinet framework, and to fasten drawer fronts and sides together.

The house should be broom-cleaned before starting to do the cabinet-work, and kept that way.

*Cabinet Parts.* The cabinet parts illustrated in Figure 24 (with the exception of partitions, which cannot be seen) pertain to a single cabinet which could be constructed as a *separate* unit prior to installation. The same terminology applies to a *built-in* cabinet, the difference being that shelf cleats would be nailed to the plastered wall; no end jambs would be used. The term "carcase" includes all parts except the front, doors, and drawers.

**Note:** The left hand door shown in Figure 24 represents a flush door (plywood). Only *one* type door would be used on a cabinet, either flush or panel. Sometimes the flush door is rabbetted to match a lip drawer. See Figure 32.

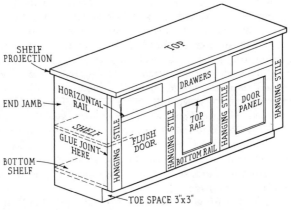

**Figure 24.** Cabinet parts.

*Cabinet Rod Layout.* A cabinet has three main dimensions: width, the measurement of the cabinet when viewed from the front; height, the distance from the floor to the top of the cabinet or, in the case of a built-in ironing board, for instance, the distance from the bottom to the top of the cabinet; and depth, the distance from front to back. The width dimension is often the largest.

The construction of a built-in cabinet (or any cabinet) is greatly simplified if a rod layout is made first. The rod need be only a piece of 1 by 2 straight-grained stock a little longer than the largest dimension of the cabinet to be constructed. The rod should be smoothed to facilitate marking the various lines. Each of three edges of the rod is used to represent one dimension of a cabinet. The fourth side of the rod is also used when a cabinet has more than one depth dimension; a cabinet may, for example, be 24 inches deep on the lower part and 12 inches deep on the upper part. Cabinets of this type are often found in a kitchen, where the wide top of the lower part serves as a work shelf.

Figure 25 illustrates the rod-layout method. Note that the depth, height, and width dimensions of the cabinet in the illustration are laid out on different faces of the rod, identified for clarity as *A* for the width, *B* for the height, and *C* for the depth. As soon as the layout is completed, each face of the rod should be labeled to identify the part it represents.

Any one of the three dimensions can be laid out first; the depth dimension, because of its simplicity, will be described first. Proceed as follows.

1. Lay the rod on a suitable surface—a bench top, for instance. Be sure that the rod is reasonably smooth.

2. Using side *C* and starting from the left end (with the rod lying in front of you), measure the depth of the cabinet, in this case 24 inches. Square a line at this point to represent the back of the cabinet.

3. Measure 1 inch net from the end of the rod to represent the front of the cabinet. Cabinet doors and drawers are fitted into a cabinet front. One inch net is the standard thickness for front material.

4. Read the blueprint and see if the cabinet is to have a back, which is usually made of ¼-inch plywood.

5. Keep the back at least 1 inch in from the back edge of the cabinet, and mark on the rod as shown. The reason for the back inset is to allow enough exposed jamb stock to scribe against a plastered wall.

Step 5 completes the depth layout. The exact width of the shelves to be fitted between the front and back of the cabinet can now be determined.

To make the height layout, proceed as follows:

1. Using a second face of the rod (side *B*) and starting from the left end, measure the height of the cabinet and square a line on the rod at this point.

2. Measure down from this mark an amount equal to the thickness of the top, which is sometimes made of wood and sometimes of tile.

3. Measure 3 inches *up* from the bottom to represent the toe space. This mark also represents the lower side of the bottom shelf.

4. Measure ¾ inch up from the toe-strip mark to represent the top face of the bottom shelf.

5. Measure 1 inch down from the underside of the top shelf to represent the width (or thickness) of the face strip placed directly under the

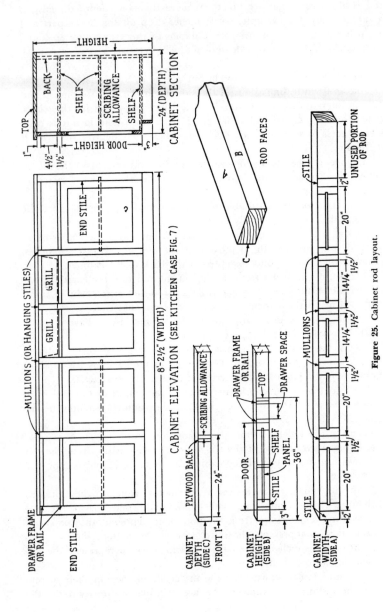

**Figure 25.** Cabinet rod layout.

top shelf. (This piece can actually be any width desired; 1 inch is used here to conform to the dimensions in Figure 25.)

6. Measure from this mark a distance of 4½ inches, which is the depth of the drawers. Square a line at this point. (A drawer has width, depth, and run measurements; the depth measurement is a vertical one.)

7. Measure 1½ inches down from the drawer line, to allow for the cross rail which separates the drawers from the doors. Square a line at this point. This line represents the top of the cabinet doors.

8. The distance from the top of the door to the lower side of the bottom shelf is the length of the cabinet doors.

Step 8 completes the height layout. To make the width layout of the cabinet shown in Figure 25, proceed as follows:

1. Using side *A* and measuring from the left end of the rod, mark a distance of 8′2½″ to represent the width of the cabinet.

2. In successive steps measure 2 inches for the left-end stile, 20 inches for a door, 1½ inches for a mullion, 20 inches for a door, etc., continuing (using the dimensions given in the illustration) until you reach the 2-inch stile on the right end of the cabinet. This completes the width layout.

The rod is now ready to be used as the basis for (1) making a cutting list of the various parts required to construct the cabinet, (2) determining the net width and length measurements of the jambs, partitions, and shelves, and (3) locating the dadoes required to support the shelves or hold the partitions.

**Note:** Cabinet doors are normally ordered from the mill as a complete unit; they should be ordered as soon as the rod layout is completed, even before the cabinet construction is started, to avoid any unnecessary delay in their delivery.

When cabinets are constructed on the job, the work includes making the drawers. Grooved drawer stock is bought at a mill; sometimes, in making the rod layout, the drawer sizes are varied slightly from those shown on the blueprint, so that the sizes will conform to the dimensions of the drawer stock.

The front material is usually not listed by separate pieces but is ordered in long lengths and then cut as required when the cabinet is constructed.

***Ordering and Selecting Cabinet Lumber.*** The exposed parts of a cabinet are made (1) of lumber that matches the wood used for the trim and doors or (2), if the cabinets are to be painted, of lumber that has good working and painting qualities. Hard or cross-grained lumber is not suitable for cabinetwork.

Either white cedar or white pine is considered excellent for the exposed parts of a cabinet which is to be painted. Birch is used for kitchen cabinets in the more pretentious homes. This wood is filled, shellacked, and varnished to retain the beauty of the grain. Sometimes selected shelving or knotty pine is ordered in long lengths, and all exposed parts, such as end jambs and cabinet tops, are made by carefully selecting the parts of a board which are free from knots. The lumber used on the inside of a cabinet is often made from shelving pine. Small tight knots are acceptable.

**Note:** At no time should lumber be used that contains loose knots; a rather sharp and dark line around a knot is evidence that it will later fall out. A tight knot blends into the surrounding wood and, if not too large, is not objectionable (see Figure 26). All knots should be treated with a coat of clear shellac, prior to painting, to prevent them from showing through the finished coats of paint.

**Figure 26.** (*A*) Encased knot; (*B*) intergrown knot.

Clear lumber should be ordered to make the cabinet-front stiles and rails. Drawer lumber is also ordered clear for the fronts, sides, and backs, although waste lumber from the cabinet carcase can often be utilized. Drawer bottoms are usually made from ¼- or ⅜-inch plywood.

***Gluing Cabinet Stock.*** A cabinet part that requires a piece of lumber more than 11 inches wide must ordinarily be glued up. End jambs that are 22 inches in width are made from two 11-inch (1 by 12) boards, or possibly from two 1 by 8 boards and one 1 by 10 board. Boards of any width may be glued together to make the required width.

Hot glue should not be used in an open room as it will congeal too quickly; hence cold glue is best for on-the-job cabinetwork. Wood wedges are usually substituted for carriage clamps as a means of holding the boards together, for the clamps are too heavy to be carried around conveniently.

**Note:** There are several standard makes of cold glue on the market. A waterproof casein glue is very good; a hardware store should be able to recommend a good brand.

All stock to be glued should be cut to rough lengths, say 1 inch longer than the net length (as determined on the layout rod). The shorter the length of stock, the easier it is to glue.

A *spring* joint should be made on the edges of the two boards to be glued together. This joint is made with a slight opening in the center, about the thickness of a thin saw blade. When pressure is applied to the center of the board to close up this space, extra pressure is automatically applied to the ends of the board. The ends of a glued board tend to open first because of air getting in (assuming a perfect joint in the first place); hence increased pressure on each end of the board prevents the glue joint from opening up.

It is essential that boards be selected that have no short kinks and that are not warped. Keeping the face of each board flush with the other face eliminates much planing (cleaning) after the boards are glued.

Planing a glued board is best done by *traversing*, a carpentry term which merely means planing across the board on a 45-degree angle. This takes off the high spots and evens the board quickly. Then plane the board with the grain until a smooth finish is secured, sandpaper lightly with a medium sandpaper, and finish with a fine paper.

Gluing wedges are made as illustrated in Figure 27. Nail several straight short pieces of lumber to the floor, keeping one edge on a perfectly straight line. Use as many blocks as necessary, the number depending on the length of the board. Make a second set of blocks in the form of wedges, using wide, short pieces of lumber and ripping and planing them on a slight diagonal, as shown. Then, keeping each pair of wedges together, as *A-A*, *B-B*, etc., nail one of each pair on the floor with enough space between these blocks and the first set to receive the stock to be glued. After the glue is applied, place the boards between the blocks, and lightly hammer the wedges until the glue joint is tight. It may be necessary to tack the glued stock to the floor to prevent it from warping while the pressure is being applied.

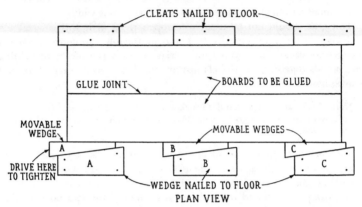

**Figure 27.** Gluing cabinet stock.

To summarize, proceed as follows when gluing up cabinet stock on the floor:

1. Select a clean space on the floor; be sure no subfloor nails are protruding; sweep this space very clean with a broom.

2. Prepare wedges and blocks.

3. Determine the location of the blocks and wedges and nail them to the floor, using 8d box nails and allowing the heads to protrude slightly for easy removal.

4. Using a jointer, carefully plane the edges to be glued. Test the joint for squareness, and be sure it is slightly open in the center.

5. Apply the glue to both edges, using a chisel-pointed stick which will fit into the glue can. The end of the glue stick should be quite thin.

6. Rub the edges of the boards together to spread out the glue evenly. This results in what is known as a *rubbed joint.*

7. To prevent the boards from sticking to the floor, lay pieces of newspaper under the joint line. The paper will stick to the board but may be removed easily.

8. Place the boards in the floor "clamp"; tighten the wedges; tack the boards lightly to the floor; and continue to tighten the wedges until a perfect glued joint is obtained.

9. Remove all excess glue; this is important, for the glue is very easy to wipe off while it is fresh but hard to remove after it is set.

10. Let the glue set a sufficient time to harden properly (at least 24 hours).

11. Remove the boards from the clamp; plane and sand smooth.

***Cutting Cabinet Parts to Exact Width.*** The exact finish width of a jamb or shelf or partition can be measured very accurately from the cabinet layout rod. Ordinarily, exposed jambs are increased in width at least 1 inch to allow sufficient lumber to make a scribed joint against the plastered wall to which the cabinet is to be fitted (see Figure 25). Shelves and partitions can be cut to their exact width, which is shown on the rod as the distance from the inside face of the front of the cabinet to the inside face of the back. If no back is required, the shelf width is the distance from the inside of the front to the finish back edge of the exposed jamb (less the scribing allowance).

***Cutting Cabinet Parts to Exact Length.*** Using the rod, determine the exact length of each piece required to make the cabinet carcase. Be sure to allow for sufficient stock to enter the various dadoes. To illustrate, if the distance between two opposing jambs is 5′4″, the shelf must be cut 5′4½″ to allow for two ¼-inch-deep dadoes.

***Dadoing Jambs and Shelves.*** Cabinets built on the job are best constructed by making dado joints into which the shelves and partitions will fit (see Figure 28). This type of joint makes the assembly of a

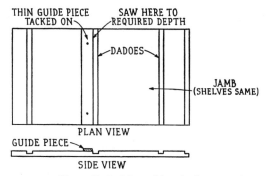

Figure 28. Dadoing cabinet jambs.

cabinet relatively easy, provided that the dadoes have been carefully laid out and cut. The layout rod is indispensable in this operation, as it shows the exact location and size of each dado.

A sharp hard pencil, steel square, backsaw, ¾-inch chisel, router, cabinet hammer, and two C clamps are required to cut a dado correctly. Proceed as follows:

1. Carefully mark each dado from the layout rod. The space *between* the dado lines must be exactly the same as the thickness of the shelving stock if a tight dado joint is to be achieved.

2. Gauge a ¼-inch mark on each edge of the jamb (for depth of dado).

3. Clamp the jamb to the bench to hold it firmly while sawing the dado cuts.

4. Select a thin piece of finish stock (¼ or ⅜ inch thick) about 1½ inches wide and 24 inches long. The length dimension will vary according to the width of the board to be dadoed.

5. Tack the guide strip exactly parallel to one of the dado lines, and, using the backsaw, make a ¼-inch-deep cut.

6. Remove the strip, set it parallel to the next dado line, and repeat the cutting process.

7. Using a ¾-inch chisel, rough out the wood between the saw cuts. Work in from each edge of the board to prevent splintering.

8. Set the router plane to the depth required, in this case ¼ inch, and smooth-cut the bottom of the dado. Be sure to work in from each edge.

9. Repeat steps 1 through 8 for each dado required.

10. To permit easy assembly, bevel each corner of a cabinet part that is to fit into a dado.

11. If a back is required and the end jambs are exposed, rabbet the back edge of the jambs to receive the back (otherwise the back would show). If 1 inch is added to the width of the jamb for scribing purposes, the width of the rabbet will need to be 1 inch plus the thickness of the back. A ¼-inch back will therefore require a 1¼-inch rabbet.

12. Clean off all surface marks on all exposed surfaces; a bottom shelf, obviously, needs cleaning only on one side.

13. Joint all exposed edges and round them slightly with sandpaper.

**Note:** The one exception to this rule is the front edge of an exposed jamb. This edge should be kept sharp and square to receive the front stile, which is glued and nailed to place to make a perfect joint.

***Assembling a Cabinet Carcase.*** Before the front of a cabinet can be cut, fitted, and fastened in place, the cabinet frame, or carcase, must be assembled. Assuming that all dadoes have been made and all exposed surfaces planed and sanded, proceed as follows:

1. Lay pieces of finish lumber, or a piece of plywood, on the floor, and place one of the end jambs on top of these pieces.

2. Select the top and bottom shelves and stand them in a vertical position.

3. Lay one of the outside jambs on these two shelves and carefully fit the dadoes over the shelves. Hammering should be done on a block of wood to prevent marring the surface of the jamb.

4. Using 8d finish nails, spaced about 6 inches o.c., nail through the face of the jamb into the ends of the shelving. If the jamb is *not* exposed, use 8d box nails.

5. Turn the shelves and jamb over onto the clean lumber on the floor, and nail on the other end jamb.

6. Fit and nail in partitions. The box nail can be used through the bottom shelf; if the top shelf is to be exposed (that is, not covered with tile or similar material), 8d finish nails must be used.

7. Square the carcase, brace temporarily, and nail on the back. A 3d box nail, spaced 4 inches o.c., is the correct size to use.

The cabinet is now ready for the front stiles and rails which will form the door and drawer openings.

*Fitting and Fastening the Cabinet Front.* Figure 29 illustrates a very simple cabinet front, composed of two stiles (vertical members), three cross rails (horizontal members), and one muntin, which forms the division between the drawers. The dimensions of the stiles and rails will vary. A standard-width stile is 2 inches net; standard-width rails are, bottom, 3 to 3½ inches; top, 1½ to 2 inches; cross rails and muntins 1 to 1½ inches. Cabinets that are constructed with a toe space do not have a front bottom rail.

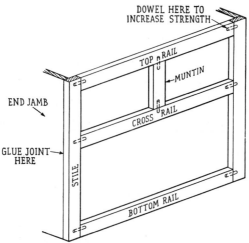

**Figure 29.** A cabinet front.

The thickness of the front stock varies from $\frac{3}{4}$ to 1 inch net. Cabinet doors which are fitted into the frame are $\frac{1}{16}$ inch thinner than the front material. On mass-housing projects many doors are rabbeted on all four edges to fit *against* the front and not into the frame. In this case the front material is $\frac{3}{4}$ inch thick.

On an ordinary on-the-job built-in cabinet the face frame materials are merely nailed to the jambs and shelves. In better construction the rails are doweled to the stiles (see Chapter 4, page 51). The stiles, which should be wide enough to extend at least $\frac{1}{16}$ inch over the face of the jamb, are glued and nailed to the exposed jambs to make a tight joint that will not open. The "overwood" is planed flush with the face of the jamb.

All inside edges of rails and stiles should be planed smooth prior to assembly.

The front is fastened to the carcase with 6d and 8d finish nails, spaced to suit the length of the stiles and rails.

After the face frame is assembled and fastened to the jamb, it should be cleaned and sandpapered. Then all exposed corners should be lightly sandpapered to remove the sharpness.

To fit and assemble a cabinet front, proceed as follows:

1. Cut all stiles and rails to rough lengths.

2. Rip and plane all stock to the net widths, as shown on the layout rod. Be sure to increase the width of the end stiles $\frac{1}{16}$ inch to allow for planing after they are glued to the exposed end jambs.

3. Cut one end of each piece square, using a miter box whenever possible.

4. If dowel joints are to be used, cut each piece, either stile or rail, to the net length, as shown on the layout rod; lay out and bore the dowel holes. Also cut in the drawer guide seats (see page 318).

5. Assemble the parts that are doweled, holding them together with carriage clamps, if available, or with wood wedges and blocks which have been made to fit the front.

6. Glue the edge of one of the end jambs and nail on one of the stiles, being sure it projects a little beyond the face of the jamb. Repeat this process for the other end jamb.

7. Cut the rails to net length and nail to place. Use glue whenever possible.

8. Plane off the overwood on the stile.

9. Lightly sand all exposed corners of all stiles and rails. This completes the cabinet front.

***Drawer-guide Construction.*** One of two types of drawer guides may be used, center or side. The center guide (see Figure 30*A*) is considered the best. It is made from hardwood and consists of (1) two pieces which have been milled with a tongue-and-groove joint or (2) a grooved guide piece into which a square runner is fitted. The square runner or the grooved part of the T-and-G type of runner is glued to the underside of the drawer bottom. The matched piece is fitted and fastened into the cabinet carcase.

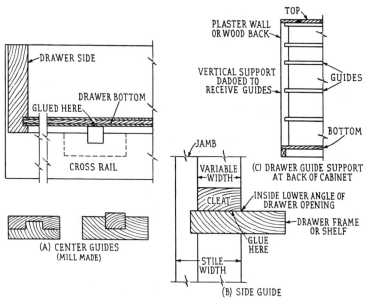

**Figure 30.** Typical drawer guides.

**Note:** Wide drawers require two center guides, placed a few inches in from each drawer side.

If a side guide (see Figure 30B) is used, a flat piece of lumber must be fitted behind the cross rail of the front, exactly flush with the top edge of the rail, to serve as a drawer runner. A second piece of lumber, nailed on top of the flat runner, acts as the side guide. The two pieces should be assembled before fastening them into the cabinet.

The key to drawer-runner assembly is to make sure that each runner is placed at exact right angles, vertically and horizontally, to the face of the cabinet. This is very important, for otherwise the drawer front will not fit the cabinet front when the drawer is closed.

The front end of a drawer runner is fastened to place by nailing through the cross rail. If the rail is wide enough, a seat should be chiseled out of the rail to receive the drawer-runner end and make it self-supporting. Nails are then required only to hold the runner to its place.

The inside end of a drawer runner is supported by a vertical piece which has been dadoed to receive each drawer runner and hold it firmly in place (see Figure 30C). This vertical piece must be securely fastened either to the cabinet back or to the plastered wall, and it must be plumb.

To fit and fasten drawer guides, proceed as follows:

1. Determine the type of guide to be used, center or side.
2. Determine how many guides are required for each drawer.
3. Prepare a vertical guide support, making the dadoes exactly match the location of the cross rails in the cabinet front.
4. Cut the guide seats in the cross rails.

Note: This operation should be done before the cabinet front is assembled; otherwise the cross rail may be damaged by the chiseling; also, there may not be sufficient room to work inside the cabinet.

5. Cut the drawer guides to exact length.
6. Assemble each guide in place, using glue whenever possible.

*Making Drawers.* The parts of a drawer are illustrated in Figure 31. Each drawer requires a front, two sides, a back, and a bottom. The measurements of a drawer are indicated as width, depth, and run.

Drawers are classified as either flush or lip. The flush drawer fits into the front members which form the drawer opening; the lip drawer has a rabbeted front which closes against the front and, at the same time, covers the space or crack around the drawer. The lip drawer is the simplest one to make and fit.

A flush or lip drawer is constructed by making a rabbeted joint in each end of the front and fitting the sidepieces into these rabbets, as shown in Figure 32. Drawer-front stock is bought from the mill already grooved and rabbeted on the edge.

Lip drawers are fitted so that they have about $\frac{3}{16}$-inch side clearance and $\frac{1}{16}$-inch vertical clearance. Flush drawers must be fitted as close as possible, depending on whether the job is to be painted or stained and varnished. The joint is made slightly larger for a painted cabinet.

Most drawer-stock cutting can be done in a miter box. All measurements must be accurately made and followed. The end rabbets on the front of a lip drawer are made $\frac{3}{8}$ inch more than the thickness of the drawer side. To illustrate, a $\frac{5}{8}$-inch drawer side requires a 1-inch rabbet on each end of the front. The standard rabbet on the top edge of the

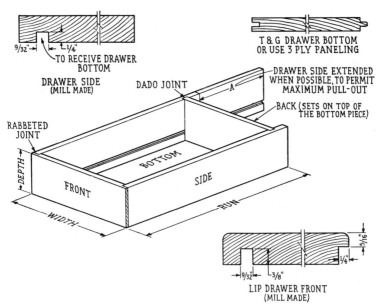

**Figure 31.** Drawer parts.

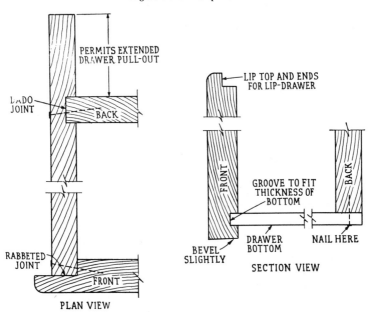

**Figure 32.** Typical drawer joints.

319

drawer front is ⅜ inch. No rabbet is made on the lower edge, for the projection of the drawer beyond the cabinet completely hides the crack.

All drawers should be constructed to permit a maximum pull-out. If the cabinet is deep enough, the back is dadoed several inches in from the ends of the sides. This permits the drawer to be pulled out all the way, since the extra side stock holds the drawer from tipping or falling out.

The width of the back is made the same as that of the drawer sides, *minus* the groove, so that the bottom can be pushed into the groove after the drawer front, sides, and back are assembled. The bottom is held in place by nailing to the lower edge of the back.

To construct a drawer, proceed as follows:

1. Measure the cabinet openings for drawer width, depth, and run measurements.

2. Select drawer stock of the correct size and rough-cut to length.

3. Lay out the drawer front (either flush or lip) and mark for the rabbets on each end.

4. Saw the rabbets with a fine backsaw, making the cuts square and accurate. Round the ends of the front to match the top edge as machined at the mill.

5. Slightly bevel the underedge of the front.

6. Determine the length of the drawer sides and cut two pieces this length. Be sure that the end cuts are square.

7. Locate the position of the back, and lay out and cut two dadoes to fit the thickness of the stock.

8. Cut the back to length, being sure to allow for the depth of the two dadoes.

9. Cut the plywood bottom according to the measurements taken from the assembled drawer. Be sure to cut the bottom perfectly square, for an out-of-square bottom means an out-of-square drawer.

10. Glue and nail the drawer together, preferably using 6d box nails or 6d cement nails and spacing them not more than 2 inches o.c. The number of nails per joint depends on the depth of the drawer. The bottom need be nailed only through the back; 3d box nails are adequate for this purpose.

11. Lightly sandpaper all exposed corners.

12. If a center drawer guide is used, lay out and mark the location of the guide on the underside of the drawer bottom. Be sure it is exactly at right (90-degree) angles to the front.

13. Glue and brad the guide to the drawer.

14. Try the drawer in the hole and adjust where necessary.

15. Rub paraffin on the guide to make the drawer operate smoothly.

*Making Shelves Adjustable.* When constructing a built-in bookcase, it is advantageous to make the shelves adjustable so that they can be arranged to fit books of various sizes.

The side jambs of the bookcase are dadoed to receive the head jamb and the bottom shelf, and are rabbeted on the back edge to receive a plywood back.

Shelves can be made adjustable in any of several different ways, which are outlined below in order of preference.

1. Four metal shelf-fixture strips, as illustrated in Figure 33*A*, are screwed in the four interior corners of the bookcase. These metal fixtures are bought in foot lengths, and cut to fit in between the head jamb and bottom shelf. Four small metal shelf rests, which fit into specially prepared holes, support one shelf. Each shelf is cut to length to fit loosely between the fixture strips. To adjust a shelf, move one set of shelf rests to the desired location.

2. A series of $\frac{1}{4}$-inch holes are bored 1 inch in from each edge of each jamb, as shown in Figure 33*B*. Holes are spaced 1 or 2 inches apart, as desired. The top and bottom holes need not be closer than 6 inches from each end of the jamb. Small metal rests, formed with a metal dowel, fit into these holes, four of them serving to support one shelf. To adjust a shelf space, move four shelf rests.

3. If metal brackets are not available, an adjustable shelf-cleat strip can be made by boring a series of 1-inch holes in the center of a selected $\frac{3}{8}$- by $1\frac{1}{2}$-inch strip of softwood (see Figure 33*C*). The length of the strip is the vertical inside dimension of the bookcase. After the holes are bored, rip the strip in two, making two cleat strips with equally spaced half-round holes. These strips are screwed into the corners of the bookcase (four strips are required). Shelf cleats are prepared with rounded ends to fit into the opposing spaces. To adjust a shelf, move two cleats. The ends of each shelf must be notched to fit loosely around the vertical strips.

4. The ratchet strip, shown in Figure 33*C*, is a variation of the bored strip; the cleats are easier to make as they can be sawed to the correct bevel in a miter box. The ratchet strips are sawed by means of a backsaw.

The layout and spacing of the ratchet strips, or the $\frac{1}{4}$-inch holes, or the metal fixture strips, must be carefully done to ensure that the shelves will be level and will not rock because of uneven shelf supports.

*Fitting a Cabinet Door.* Basically, a cabinet door is hung in the same manner as an interior door. Naturally the cabinet door is much easier to handle, and it can often be held in a bench vise. If the door is to be

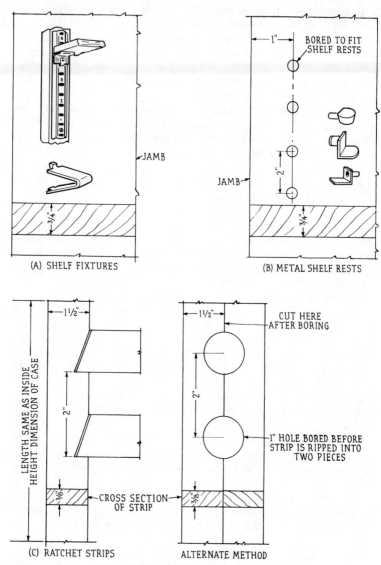

(A) SHELF FIXTURES

JAMB

3/4"

(B) METAL SHELF RESTS

1"

BORED TO FIT SHELF RESTS

JAMB

2"

3/4"

(C) RATCHET STRIPS

1 1/2"

2"

LENGTH SAME AS INSIDE HEIGHT DIMENSION OF CASE

3/8"

CROSS SECTION OF STRIP

ALTERNATE METHOD

1 1/2"

CUT HERE AFTER BORING

2"

1" HOLE BORED BEFORE STRIP IS RIPPED INTO TWO PIECES

3/8"

**Figure 33.** Adjustable shelf cleats and rests.

fitted into a cabinet which has a toe strip, the lower edge is free; that is, the door does not have to be fitted to a bottom rail. This is also true for a cabinet located above a sink since its lower edge is free.

Whether the door is to be fitted on three edges or on four, the procedure is the same:

1. Measure the width and length dimensions of the door and check with the size of the opening to determine how much will have to be planed off the door.

2. Select the hinge edge of the door. There is no set rule for this if the blueprint does not give the information. A cabinet door should swing to a wall to make it easy to reach inside the cabinet.

3. Plane the hinge edge square and straight.

4. Measure the width of the door opening; mark this measurement lightly on the door, and plane the front edge down to this line. This edge will need a slight underbevel to permit the door to open freely. After planing, try the door until the required joint is achieved. The joint should be larger for a cabinet with a painted finish than for one with a stain-and-varnish finish. A $\frac{1}{16}$-inch crack on each edge is sufficient for the paint job.

5. Place the door into the opening; check the top edge with the top of the opening and plane where needed. Replace the door in the opening and check the joint to be sure it is exactly parallel.

6. If the lower end of the opening is free, that is, if there is no bottom rail, mark the door so that it lines up exactly with the lower ends of the door stiles; then plane the door until it is exactly flush.

7. If there is a lower rail, measure the vertical distance in the opening, mark this measurement on the door, and plane. Try the door in the opening and plane as necessary to secure a joint of the correct size.

8. Lightly plane all edges of the door to remove the sharpness.

***Hinging a Cabinet Door.*** If the cabinet door is to be hung with loose pin butts, follow the instructions given on page 274. If a butterfly or surface hinge is used, proceed as follows:

1. Screw the hinge leaf to the face of the door stile. Ordinarily the top of the top hinge is lined up with the inside edge of the top rail, and the lower end of the bottom hinge is lined up with the inside edge of the bottom rail.

3. Place the door in the opening; wedge in place with very small wedges (sometimes 4d finish nails will do) and equalize all space around the door.

4. Screw the hinge leaves into the door stile.

5. Test the door, particularly checking the bevel on the front edge to be sure it clears properly as the door is opened.

## FITTING AND HANGING WINDOW SCREENS

The modern window screen is very different from the type used many years ago, which consisted of a wooden frame, made of 1 by 2 redwood stock, covered with screen wire. The wire was fastened on with wire tacks, which were then covered with a small half-round screen bead. Today, many screens consist only of the screen wire, on each end of which a small metal bar is clamped. To keep the wire taut, the top and bottom metal bars are fastened either onto a roundheaded screw with a keyhole slot or onto hooks.

If a wood-frame screen is to be hung, follow the general instructions given above for hanging cabinet doors. If an all-metal screen is used, follow the printed instructions given by the manufacturer.

## STAIR CONSTRUCTION

Stairbuilding is recognized as one of the carpentry jobs that require the highest degree of skill; often the job is done by a specialist known as a stairbuilder. There are, however, many forms of simple stairs, such as a basement stairway, that are not difficult to construct provided that one possesses a basic knowledge of tool usage and also a good degree of woodworking skill. Nevertheless, because of the complexity of the layout and actual construction of a stairway, the subject is too large to be treated here.*

## INSTALLING HARDWOOD FLOORING†

In many communities the installation of hardwood flooring is done by specialists known as floor layers. The floor layers are then followed by other specialists known as floor finishers. Floor-laying operations, however, easily lend themselves to the carpenter's skill, and the installation processes are therefore described below. The discussion is limited to oak flooring, but *the same general rules and procedures will apply* to maple and beech hardwood flooring.

* Readers interested in doing their own stairbuilding should see J. Douglas Wilson and S. O. Werner, "Simplified Stair Layout," Delmar Publishers, Albany, N.Y., 1947, or E. A. Lair, "Carpentry for the Building Trades," 2d ed., McGraw-Hill Book Company, Inc., New York, 1953.

† Much of the following information has been taken or adapted from "Oak Floors for Your Home," a pamphlet issued by the National Oak Flooring Manufacturer's Association. Permission to use the excerpts is hereby acknowledged with appreciation.

For purposes of clarity, the descriptive information on oak flooring is divided into eight sections: (1) Types, Sizes, Selection, and Grading; (2) Proper Installation Time; (3) Preparation before Installation; (4) Applying Building Paper; (5) Arrangement of Flooring Pieces; (6) Adequate Nailing; (7) Laying Procedures; and (8) Sanding and Finishing.

*Types, Sizes, Selection, and Grading.* There are three general types or styles of oak flooring: strip, plank, and parquet. Only the strip type will be discussed, as it is the most economical and hence the most popular.

Strip flooring, which consists of flooring pieces cut into narrow strips, is available in several different thicknesses and widths. It is always laid in random lengths; that is, the end joints are scattered and not clustered.

Despite their extensive use, strip floors retain individuality in character and beauty of grain. No two oak floors are exactly alike. Most strip floors are composed of pieces of uniform widths, but interesting effects can also be achieved by using random or mixed widths. Interesting patterns, in either case, can be attained by using pieces selected for variations in color, mineral streaks, or other natural irregularities.

Most strip flooring is tongued and grooved at the flooring mill. A special machine cuts a tongue on one side and end and a groove on the other side and end (see Figure 34). The tongue and groove allow the flooring pieces to join one another more snugly than would otherwise be possible.

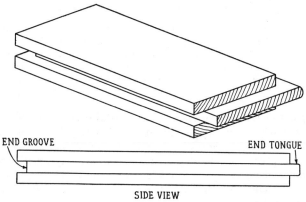

END GROOVE            END TONGUE

SIDE VIEW

**Figure 34.** Side and end T-and-G matching.

Another feature of modern strip flooring is the undercut. This consists of a groove on the bottom of each piece which enables it to lay flat even though the subfloor surface may contain slight irregularities.

The grading of hardwood flooring is based primarily on appearance, for all regular grades of oak flooring possess adequate strength, durability, and resistance to wear. Chiefly considered are such characteristics as knots, streaks, pinworm holes, the amount of sapwood, and variations in color. The most commonly used grades of quartersawed oak, are, in descending order, clear, sap clear, and select. Plain-sawed grades are clear, select, No. 1 common, and No. 2 common. Figure 35 illustrates the difference between quartersawed and plain-sawed oak. Quartersawed oak is considered the most desirable although it costs more.

Oak strip flooring is manufactured in several sizes, ranging in width from $1\frac{1}{2}$ to $3\frac{1}{4}$ inches and in thickness from $\frac{5}{16}$ to $\frac{25}{32}$ inch. Standard thicknesses for the tongue-and-groove styles are $\frac{1}{2}$ inch, $\frac{3}{8}$ inch, and $\frac{25}{32}$ inch. The $\frac{25}{32}$-inch pieces, which are employed most extensively, are available in four widths: $1\frac{1}{2}$, 2, $2\frac{1}{4}$, and $3\frac{1}{4}$ inches. The $2\frac{1}{4}$-inch width is the most popular in most sections of the country. Square-edged strip flooring of oak is ordinarily produced in two sizes, $\frac{5}{16}$ by $1\frac{1}{2}$ inches and $\frac{5}{16}$ by 2 inches.

**Proper Installation Time.** The installation of hardwood or finish flooring should be the last construction operation in a house. All plastering, plumbing-fixture installations, electrical wiring, and electrical-fixture installations, all trim and cabinetwork, painting, and paper hanging, should be completed before the finish floor is begun. The purpose of this procedure is to ensure that the finish flooring will not be used as a working surface by the building mechanics. This is very important.

Note: Disregarding these instructions devised by oak-flooring manufacturing experts, some contractors will lay the flooring before the interior finish work is done, on the theory that all twisting, shrinking, and warping will take place prior to sanding. The flooring material should be delivered at least five days prior to laying, and stacked in loose bundles to permit its adjustment to the moisture content of the new structure.

**Preparation before Installation.** A beautiful oak floor is eye-catching and often reflects the quality of construction that has gone into the house, for obviously a beautiful floor would not be installed in a house where cheap construction was the main objective. If a floor is to stay beautiful and not warp or shrink or swell, the problems of moisture, ventilation under the floor, and the condition of the subfloor must be carefully checked.

VISIBLE MOISTURE. Visible moisture is in evidence in several ways:
1. Water in the basement or understructure of the building.
2. Leaks in plumbing, radiators, roofs, and walls.
3. Rain coming through open windows or doors.

White oak, quartered

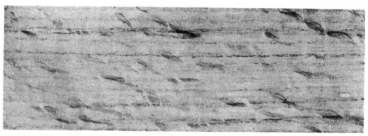

Red oak, quartered

White oak, plain

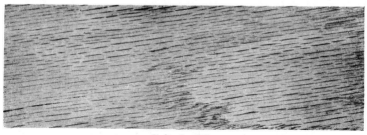

Red oak, plain

**Figure 35.** Plain- and quartersawed oak.

The correction of these conditions *before* the floor is installed will go far toward ensuring trouble-free service from floors.

INVISIBLE MOISTURE. Invisible moisture is harder to check; it may result from any one of several conditions:

1. Excessive atmospheric humidity in the area or locality where the floor is installed. This hazard can be minimized by leaving the proper expansion space at the walls where the hardwood flooring is started and finished (at least ¾ inch should be allowed between the face of the wall and the flooring strips).

2. Plaster that is not thoroughly dry. The plaster should be tested by placing the hand flat against the wall. If the wall feels damp, the plaster is not dry.

3. Concrete foundation walls that still retain moisture. This condition exists when too much speed was used in getting the frame built on top of the foundation, so that the concrete was not first allowed to become thoroughly dry.

4. The use of green lumber for subflooring and floor joists.

The basic rule to follow when some of these invisible-moisture conditions obtain is to delay installing the floor as long as possible.

VENTILATION. Ventilation is highly essential in order to prevent wood floors from being damaged by moisture. An adequate cross circulation of air under the building can be attained by providing vents in the foundation walls. Vent openings should total in area at least 1½ per cent of the floor area of the first floor. Before flooring is installed, the building should be ventilated, on clear days, whenever possible. It should also be ventilated periodically after all work is finished and the building is awaiting occupancy.

SUBFLOORS. Building experts are unanimous in declaring that, in conventional joist construction, *sound* subflooring under the finish flooring is virtually a "must." Ordinarily nailed directly to floor joists, subfloors are so essential to good construction that most modern building codes specify them. Their omission is usually poor economy, even where the finish floor is to be of strong, durable oak.

Subflooring serves several important purposes. It lends bracing strength to the building and provides a solid base for the finish floor, practically eliminating the possibility of floor sag and squeaks. By acting as a barrier to cold and dampness, it helps keep the building warmer and drier in winter. In addition, it provides a safe working surface during the erection of the house.

Subflooring in new buildings should be constructed of 1 by 6 ordinary lumber stock, No. 1 or No. 2 common grade, thoroughly seasoned and dry. Boards should be square-edged, never wider than 6 inches. Each board should be face-nailed at every bearing with two 8d nails, and a

¼-inch expansion space should be left between boards. All butt joints must rest on bearings. Subflooring should be laid diagonally with the joists at a 45-degree angle.

In remodeling, where hardwood flooring is laid over old wood floors, all conditions applicable to new wood subfloors should apply. All rotten, splintered, weakened, or worn boards should be replaced. The entire area should be examined for loose and/or weak spots, and any defective places renailed to ensure a solid foundation for the new floorings.

Just before installation of the finish flooring is to begin, the subfloors should be examined carefully and any defects corrected. Raised nails, for instance, should be driven down and loose or warped boards replaced. Then the subfloors should be swept thoroughly and, if necessary, scraped to remove all plaster, mortar, or other foreign materials. These precautions must be observed if the finish flooring is to be laid properly.

*Applying Building Paper.* The last operation before the actual laying of the oak finish floor is to apply a good quality of building paper over the subflooring. This will protect the finish floor and the interior of the house from the dust, cold, and moisture which might otherwise seep through the floor seams. In the area directly over the heating plant, it is advisable to use double-weight building paper or standard insulating board. This will protect the finish floor from excessive heat, which might otherwise cause the floor boards to shrink.

*Arrangement of Flooring Pieces.* In most houses strip flooring presents the most attractive appearance when laid lengthwise of the longest dimension. In some cases, however, it is considered acceptable to lay the flooring crosswise of the building. This is true only if the rooms are sufficiently wide. Proper placing of the strips of flooring calls for the use of the shorter lengths in closets and in the central areas of the less important rooms and of the longer pieces at entrances and in the border areas of the main rooms. The floor as a whole is most attractive when this arrangement is used. Care should be taken to stagger the end joints of the flooring pieces (a minimum of 12 inches should be left between them) so that several are not grouped closely together.

*Adequate Nailing.* Adequate nailing is absolutely essential in the finish floor, as well as in the subfloor. Insufficient nailing may easily result in loose or squeaky floors, one of the most annoying deficiencies a house may develop. Tongue-and-groove flooring should be blind-nailed: that is, the nails should be driven at an angle of about 50 degrees at the point where the tongue leaves the shoulder (see Figure 36). In blind nailing, the head of each nail is countersunk with a steel nail set. Whatever the type or size of flooring used, the mechanics should closely follow the nail schedule recommended by the manufacturer.

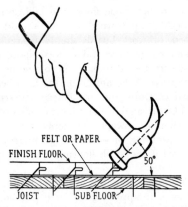

**Figure 36.** Angle of nail when blind-nailing T-and-G hardwood flooring.

*Nailing Specifications.* Follow the specifications in Table 1 in nailing all tongue-and-groove floorings.

Table 1. Nailing specifications (over wood subfloors)

| Size of flooring | Size of nail | Spacing |
|---|---|---|
| $2\frac{5}{32}'' \times 3\frac{1}{4}''$<br>$2\frac{5}{32}'' \times 2\frac{3}{4}''$<br>$2\frac{5}{32}'' \times 2\frac{1}{4}''$<br>$2\frac{5}{32}'' \times 1\frac{1}{2}''$ | 7d or 8d cut-steel flooring or screw-type nails | 10″ to 12″ o.c. |
| $\frac{1}{2}'' \times 2\frac{1}{2}''$<br>$\frac{1}{2}'' \times 2''$<br>$\frac{1}{2}'' \times 1\frac{1}{2}''$ | 5d or 6d cut-steel flooring or screw-type nails | 8″ to 10″ o.c. |
| $\frac{3}{8}'' \times 2''$<br>$\frac{3}{8}'' \times 1\frac{1}{2}''$ | 4d cut-steel flooring or screw-type nails | 6″ to 8″ o.c. |

**Note:** Sufficient nailing is important. It makes the floor solid, thus helping to prevent squeaks, and tends to retard expansion.

*Laying Procedures.* In a new building, all baseboard, trim, and door-jambs should be set so as to allow free movement of the flooring beneath them.

Finish flooring should be laid at right angles to floor joists, and should run the long way of the room wherever possible. Where subflooring has not been installed diagonally across the joists or where new flooring is being laid over old surface floors, the new floor should be laid at right angles to the existing floor or subfloor.

Flooring should run continuously between adjoining rooms. This procedure often results in long runs, which must be kept in a straight line in order to attain a craftsmanlike job. To secure this result, a "starter" chalk line must be made as a guide to work from, as it cannot be assumed that the "starter" walls of two adjoining rooms are in a perfectly straight line. The procedure for laying a hardwood floor, as shown in Figure 37, is as follows:

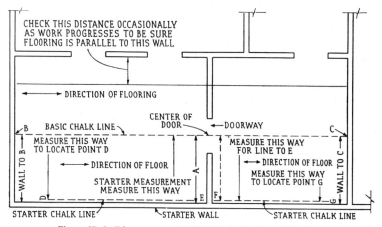

**Figure 37.** Striking a starter (chalk line) for hardwood flooring.

1. Measure from the wall from which the flooring is to be started to the approximate center of the doorway through which the flooring is to be laid (line *A*).

2. Using this figure as the basic measurement, locate points *B* and *C* as illustrated (wall to *B* and wall to *C*).

3. Strike a line from *B* to *C*.

4. Measure *toward* the wall from this chalk line (*B-C*) and strike two starter chalk lines, *D-E*, and *F-G*. These lines will then be parallel to the chalk line which passes through the doorway.

5. Occasionally measure from the edge of the front strip of flooring to line *B-C* to be sure the flooring is kept parallel. It is a demonstrable fact that when two floor layers are working on opposite ends of the same flooring run, one end may be nailed tighter than the other end, since

one worker may be applying more driving power than the other. The result will be a bowed or crooked flooring run. An occasional check from the main chalk line will enable the floor layers to keep the flooring in a perfectly straight line. This is very important, as a crooked flooring run can never be hidden, except to the extent that it may be covered with rugs.

6. Start laying flooring with the grooved edge of the first strip along a wall line, with at least a ¾-inch expansion space between the plastered wall and the flooring strip. Face-nail the first strip along the edge nearest the wall so that nailheads will be covered by the shoe molding. Blind-nail the tongue side. The first nail in each strip should be pointed toward the preceding strip in order to help eliminate cracks at end joints. Do not drive strips excessively tight together.

7. It will be necessary to use a pinch bar and block to make a tight joint between the last few strips of flooring, since there will not be sufficient room to swing a hammer. Face-nail the last board in the same manner as the first board, keeping the nails as close to the baseboard as possible so that the shoe molding will cover them.

*Sanding and Finishing.* The first of the several finishing operations is sanding, also known as surfacing. This should be done just before the final coat of finish is applied to the base-shoe molding and after all other interior work has been completed. Preparation of the floor for sanding is important. It should be swept clean, but no water should be used.

In most cases, the greatest part of the sanding should preferably be done with an electric sanding machine, since this method saves both time and money. Large machines, however, are not efficient near walls, in corners, or in small closets. Such areas, therefore, should be sanded manually or with a small, power-driven hand sander. Especially fine floors merit hand sanding over the entire surface.

The average floor should be traversed four times. On the initial traverse the sanding should first be crosswise of the grain, then lengthwise. It is best to use No. 2 sandpaper for the first traverse, No. ½ for the second, and No. 0 or No. 00 for the third and fourth. Then the floor should receive a final buffing, or cleaning, by hand, with No. 00 or No. 000 sandpaper or with fine steel wool. The latter should not be used if paste wood filler is to be applied immediately thereafter. After the final buffing, the floor should not be walked on until the stain, filler, or first coat of finish has been applied and has dried thoroughly.

# CHAPTER NINE

# Finish Hardware

Finish-hardware installation includes the application of entrance and interior lock sets (see Figure 1), window fasteners, casement-sash fasteners, cabinet doorknobs and catches, drawer knobs, clothes hooks, and a group of miscellaneous items, including push plates, door bumpers, door grilles, and screen hangers.

**Figure 1.** Interior and exterior lock sets.

Because of their limited use in ordinary residential construction, such special hardware items as a cylinder deadlock or a lock for a French window or door are not described here. Detailed printed installation instructions are always packed with an intricate piece of finish hardware, and experience gained in applying ordinary hardware items will help simplify the application of these special items.

**Note:** Door and casement-sash butts (hinges let into the edge of a door or sash) are required for hanging communicating * and exterior doors. Although they are a finish-hardware item, a description of their installation is not given here, but is included in the Doorjamb section of Chapter 8. Ornamental hinges (hinges fastened to the surface of a door) are considered an installation item since they are used in connection with the construction of a cabinet.

Hardware installation is often done in two stages: (1) fitting interior and entrance-door locks before the painting work is begun and (2) applying the door locks and surface hardware after the interior decorating has been completed. The latter operation requires especially skillful tool usage, for the carpenter can easily mar the surface of a piece of finished work by making careless layout markings or by letting the screwdriver slip when he is turning in the screws. The cutting in of the front edge of a door lock, for example, must be done very carefully if a perfect fit is to be obtained.

## HARDWARE MATERIALS AND FINISHES

A basic knowledge of hardware types, finishes, and materials is essential to an intelligent understanding of the subject of finish hardware. An experienced builders' hardwareman is the best source of information for anyone contemplating buying finish hardware to complete the interior of a house. Hence only a brief description of the types and finishes of hardware is given here; the installation of various hardware items, however, is described in more detail.

*Materials.* Finish-hardware materials are made from steel, solid brass, or bronze. The ordinary residence, except for the front-door escutcheon plates, has little solid brass. Bronze hardware is used primarily in public buildings, where maintenance is an important item, but a costly residence would also use bronze, which is the most expensive hardware obtainable. The various exposed parts of steel-plated hardware, which is most commonly used in average homes, are thin and may tarnish easily unless

* The term *communicating door* is used to differentiate the doors used for room entrances from cabinet doors.

they are properly lacquered. Hardware items made from solid brass are much thicker than the plated steel; this difference can be easily seen. Solid brass, by weight, is also heavier.

Push plates for double-acting doors are made from polished plate glass. Cabinet drawer and doorknobs are made of metal, wood, or glass.

*Finishes.* Finish hardware is manufactured in a variety of finishes, including brass, copper, sandblast, black, and nickel. Nickel hardware is most often used in the kitchen and bathroom. The type of hardware finish to select is based on the taste of the owner and on the general decorative schemes of the various rooms.

## HARDWARE INSTALLATION

*Tools.* Several different sizes of screwdrivers are essential in installing finish hardware. A long-shank screwdriver with a narrow, thin end is very useful for turning in small screws. A brace and set of bits are required for installing inside-door locks; several different widths of finish chisels are needed for cutting in lock faces and strike plates; a butt chisel (short and wide) simplifies the cutting in of a hinge leaf. A sharp pocketknife is a necessity when scoring around the edges of the faceplate preparatory to cutting it in.

A screw-centering punch is very useful in quickly and easily punching pilot holes for screws. The holes in the hinge are used to locate the position of the punch and to center the hole automatically.

A finish (12-ounce) hammer and a rule are also required.

*Spacing.* The spacing of a piece of finish hardware and the position of the slots in the screwheads are important factors in securing a pleasing finished installation job. Screw slots should always be left in a vertical position; cabinet doorknobs are placed slightly above the center of the door; drawer knobs are spaced vertically, slightly above the center of the drawer and, horizontally, in the exact center, unless two knobs are used, in which case the knobs are placed the same distance in from each end; no set rule can be given for this measurement.

Mortised door locks are centered on the edge of the door stile, with the center of the doorknobs 36 inches from the floor. Casement-sash fasteners (a surface item) are centered vertically on the sash stile.

Coat and hat hooks screwed to the hook strip in a clothes closet are spaced according to the length of the wall and the number of hooks desired; 8 inches o.c. is satisfactory.

***Butts and Hinges.*** (See Figure 2.) Butts are either loose-pin or tight-pin; the loose-pin type permits the door to be taken off easily. In giving butt sizes the length is always stated first, then the width. It is important to remember this; otherwise the butts may be ordered incorrectly, and the wrong size delivered.

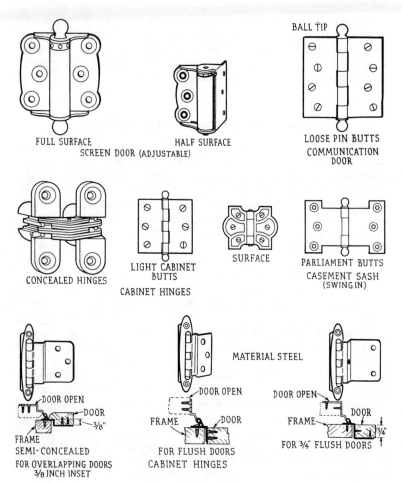

**Figure 2.** Butts and hinges.

Door butts for interior communicating doors are made in a variety of sizes, the most common size being $3\frac{1}{2}$ by $3\frac{1}{2}$ inches (the dimensions of a hinge when opened flat). An average interior door requires one pair: an exterior door requires one and one-half pairs (three butts). A standard cabinet butt is $2\frac{1}{2}$ by $2\frac{1}{2}$ inches or $2\frac{1}{2}$ by 2 inches.

Casement sash that swing out are usually hung with 3- by 3-inch butts, which are sheradized to prevent rusting. For the swing-in sash a parliament butt is required. This is a hinge with a wide enough swing to provide clearance for the window shade which is screwed to the face of the sash. Standard sizes are 4 by $2\frac{1}{2}$ inches or $4\frac{1}{2}$ by $2\frac{1}{2}$ inches (4 inches is the width of the hinge when opened).

Ornamental hinges have various names, such as semi-invisible pivot hinge, semiconcealed cabinet hinge; invisible (Soss) hinge, or simply surface hinge. These hinges are all tight-pin; hence a door cannot be removed unless the screws are removed.

Few specific rules can be given for placing a pair of hinges on a door. On interior communicating doors, however, carpenters use the rule that hinges should be 7 inches down from the top of the door and 11 inches up from the bottom. Surface hinges are often lined up with the inside edges of the door rails. Good design calls for the lower hinges to be located a greater distance from the bottom of the door than the top hinges are located from the top.

The cutting-in of a hinge seat has been described in Chapter 8, page 274. Surface hinges are screwed to the face of the door stiles and the face of the frame stiles.

*Fitting Door Locks.* (See Figures 3 and 4.) Locks for communicating doors are manufactured for interior doors, exterior doors, and screen doors. There are three general types of locks, namely, (1) the rim lock, rarely used today; (2) the mortise lock, and (3) the bored-in lock. The rim lock is screwed on the face of the door, the mortise lock is mortised into the door stile, and the bored-in lock, as its names implies, is fitted by boring holes to receive the lock cylinders and the lock latch. Because of its simplicity of installation, the bored-in type is the most popular lock in use today.

Note: A left-handed lock is required for doors which open toward you that have the butts on the left-hand side; a right-handed lock is required for doors which open toward you that have the butts on the right-hand side. A lock manufacturer cannot know which way a door will open; hence locks are made reversible: the cover of the lock may be taken off and the door latch reversed by turning it over. Extreme caution is needed for this comparatively simple operation to ensure that other parts of the lock do not fly out while the latch is being reversed.

Front-door locks are more pretentious than inside-door locks. Ornamental escutcheon plates are provided to give locks an attractive appear-

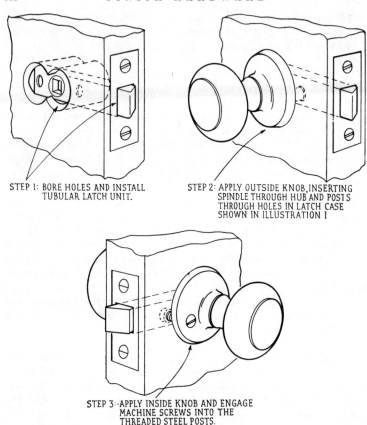

STEP 1: BORE HOLES AND INSTALL
TUBULAR LATCH UNIT.

STEP 2: APPLY OUTSIDE KNOB, INSERTING
SPINDLE THROUGH HUB AND POSTS
THROUGH HOLES IN LATCH CASE
SHOWN IN ILLUSTRATION 1

STEP 3: APPLY INSIDE KNOB AND ENGAGE
MACHINE SCREWS INTO THE
THREADED STEEL POSTS.

**Figure 3.** The three steps required to fit a bored-in lock.

ance. Escutcheon plates are often a different finish on either side of the
same door. The rule is to match the room from which the plates can be
seen. To illustrate, a door opening into a bathroom from a hall finished
in brass hardware would require brass-finish hardware on the hall side
and nickel-finish hardware on the bathroom side.

It is essential that a door be held rigid while the lock cutting is being
done. On a one-panel door, the simplest device is to drive a wedge under
the door while it is partially open. On a cross-rail five-panel door, two
smooth sticks are carefully wedged in opposing positions between the
subfloor and the lower edge of the center rail.

The standard height of a doorknob, measured from the finish floor

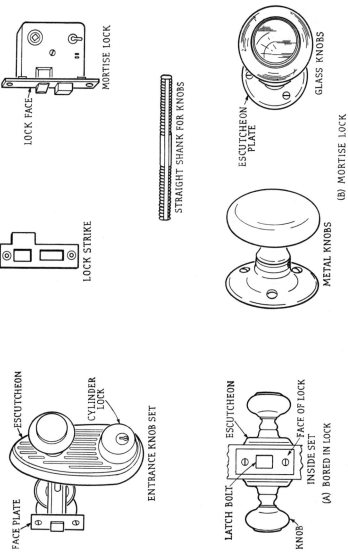

**Figure 4.** Terminology of parts of communicating-door locks.

to the center of the knob, is 36 inches. A front-door lock is made the same height as the inside doorknob, and the exterior door handle is then fitted to meet the requirements of the lock.

The layout for the knob-spindle holes must be measured according to the dimensions of the lock being fitted. Manufacturers of good locks enclose a printed layout template with the lock. Pricking through the dots marked on the card locates the center of the various holes to be bored. These layout templates are made to fit doors of varying thicknesses.

All boring or mortising must be done parallel to the face of the stile and, horizontally, must be perfectly level. Unless this accuracy is obtained, the lock will not work freely, and the finish faceplates, which are inserted in the edge of the door stile, will not be flush with the wood.

The strike plate must be carefully lined up with the door latch to ensure that the latch enters the center of the plate. The distance from the edge of the doorstop to the outside edge of the hole in the strike plate which engages the door latch must be measured very carefully. If the strike plate is too close to the stop, the door latch will not enter the strike plate; if the distance is more than required, the potential result is a door rattle.

**Note:** To prevent marring the finished surface of the door and to eliminate the possibility of getting paint on the exposed hardware parts, the door locks should be fitted *before* any painting is done, and then removed and replaced in the original box. They are reinstalled after the decorator is through with his work. During the finishing and decorating period, front doors are kept "locked" by means of a block nailed to the subfloor. The lock on the rear door is usually left in during this time to permit locking up the house.

FITTING A BORED-IN INSIDE-DOOR LOCK. (See Figure 3.) A bored-in inside-door lock is fitted as follows:

1. *Before* opening the lock box, read the description on the outside to make sure that you have the right type of lock. Some locks are made for 1⅜-inch doors; others for 1¾-inch doors, and still others for a door of either thickness.

2. Carefully open the box over a clean, smooth area, for screws may be lost as the various parts of the lock are unwrapped. Save the lock template and all printed installation instructions.

3. Examine all the parts to be sure that none is missing and that the finish of the hardware is correct; the building specifications may need checking on this point.

4. Select augur bits of the correct size to fit the lock cylinders; the size of bit to use is included in the instructions or printed on the lock template.

5. Wedge the door in a position which permits easy boring and chiseling.

6. Check the lock to see if it will fit the swing of the door without reversing the lock latch. The short side of the bevel on the latch should be on the stop side of the door.

7. Measure 36 inches from the top of the subfloor; then add the thickness of the hardwood flooring.

8. Place the lock template on the door as indicated on the printed instructions, making sure that it is at the correct height; then mark the center of all holes.

9. Remove the template, and bore the latch hole first.

10. Bore the cylinder holes from *each side* of the door. This is very important as the augur bit will splinter the face of the door stile if a hole is bored from one side only.

11. Place the door latch in place and, using a pocketknife, carefully score around the edge of the faceplate attached to the latch.

12. Using a sharp chisel of the correct width, cut in a seat for the faceplate. Because of the bevel of the door edge, the plate must be dapped in until it is exactly flush with the door edge. (This will make one edge of the faceplate slightly below the door edge.)

13. Screw the door latch to the door; then assemble the lock spindles and knobs as directed in the printed instructions. Be sure, in tightening the knobs, that the latch works freely.

14. Close the door and make a careful layout for the strike plate. Be sure to center the plate on the latch so that the door can close freely.

15. Dap in the strike plate and screw.

16. Remove the wood inside the latch holes to permit the door latch and bolt to enter the jamb. Close the door and lock it to see if it operates freely. Adjust where necessary.

17. Remove the lock and replace it in its original box. The lock is reinstalled after the interior decorator is through with the painting and decorating.

FITTING A MORTISE LOCK. (See Figure 4.) The above instructions for fitting a bored-in lock are also applicable to the mortise lock. Additional instructions are needed, however, for cutting in a mortise large enough to receive the lock, which is roughly $\frac{5}{8}$ inch thick and 4 inches square.

1. Gauge a center line, 4 inches long, on the edge of the door stile, using the knob-height measurement, 36 inches, as the basis for locating the line.

2. Make two marks on the door stile to represent the ends of the lock.

3. Bore the knob-spindle holes, working from each side of the door. Ordinarily $\frac{5}{8}$-inch holes are large enough. If the holes are too large, there will not be sufficient wood left around the holes to hold the escutcheon-plate screws.

4. Using the center line as a guide to locate the center of the augur

bit, bore a series of holes to a depth of approximately 4 inches. The lock should be measured to ascertain the exact depth required. For an average lock, use a ⅝-inch bit. The holes should be bored slightly larger than the thickness of the lock to permit it to slide freely into the mortise.

5. Using a strong-shank chisel, remove the wood between the holes, continuing until the mortise is completed.

6. Push the lock into the mortise and knife-score around the faceplate.

7. Carefully remove the wood inside the knife lines to obtain the required depth, which is the thickness of the faceplate. The door bevel necessitates that one edge of the faceplate be slightly below the face of the wood.

8. Screw the lock into place.

9. Remove one of the doorknobs from the spindle, slip on an escutcheon plate, push the spindle through the lock, put on the other escutcheon plate, and then replace the knob.

10. Carefully locate the escutcheon screw holes, being sure that the knobs are exactly level. This is done by sighting or by testing with a try square held against the face of the door. Screw on the escutcheons.

11. Locate the strike plate and cut it in as described for the bored-in lock in steps 14, 15, and 16 on page 341.

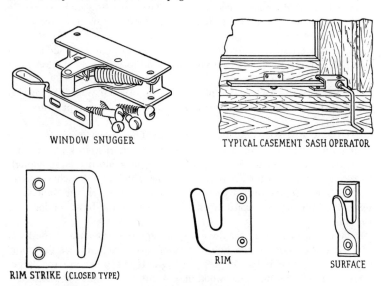

WINDOW SNUGGER        TYPICAL CASEMENT SASH OPERATOR

RIM STRIKE (CLOSED TYPE)      RIM      SURFACE

THREE TYPES OF STRIKE PLATES

**Figure 5.** Casement-sash hardware (Snugger also applicable to doors).

12. Remove the lock and repack. Label the box to indicate the door to which the lock belongs—for example, "door from living room to hall."

With the advent of the walk-in wardrobe closet, a new type of hardware has been developed. This is known as a "Snugger" (the trade name for Win-Dor No. 48; see Figure 5 *). This item consists of two parts—an encased spring and a hook. Installation is very simple as printed instructions are packed with each Snugger. The encased spring, or body, is fastened to the underside of the head jamb and against the lock edge of the jamb. The hook part of the item is carefully located and screwed to the inside face of the door in such a way that it automatically locks the door into place when the door is closed and yet permits the door to be opened very easily. A single doorknob, located at the usual doorknob height (36 inches), is all that is required to open the door.

**Note:** This hardware item has other uses, such as for large casement sash and double doors.

*Installing a Casement-sash Fastener.* (See Figure 5.) A casement-sash fastener is located, vertically, midway on the sash stile. The fastener consists of two parts, the handle and the strike plate. To install the fastener, proceed as follows:

1. Measure for the vertical center of the sash and make a light mark at this point.

2. Locate the plate which holds the movable handle so that it centers on the vertical center mark and on the stile width. When it is in this position, make pilot holes for the screws by using the self-centering screw punch.

3. Screw the fastener to the stile.

4. Carefully locate the position of the strike plate so that the end of the locking bar will enter the slot and come to a level position. Score around the strike-plate edges with a pocketknife.

5. Chisel in the strike-plate seat, being sure that the bottom of the seat is flat.

6. Screw the strike plate to position.

7. Test the casement fastener for ease of operation; the sash should not rattle when locked.

**Note:** Ordinarily casement-sash hardware is *not* removed prior to painting, as it is quite important to be able to lock the sash during the finishing and decorating period. A skillful painter can "cut" around the hardware without difficulty.

To hold a swing-out casement sash stationary, a casement operator is required. There are several types, some of which require installation at

---

* Illustration courtesy of Win-Dor Casement Hardware Co., Chicago, Ill.

the time the apron is nailed on. The simplest type, shown in Figure 5, is screwed to the face of the sash and on top of the window stool.

*Installing a Double-hung-window Fastener.* (See Figure 6.) A double-hung-window lock is screwed to the meeting rails of the upper and lower sash. The base or locking part is screwed to the top face of the lower-sash meeting rail; the strike plate is screwed to the inside face of the top meeting rail. The installation procedure is as follows:

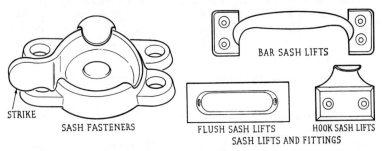

STRIKE
SASH FASTENERS

BAR SASH LIFTS

FLUSH SASH LIFTS
SASH LIFTS AND FITTINGS

HOOK SASH LIFTS

**Figure 6.** Double-hung-window hardware.

1. Measure the width of the window and divide by two to determine the center position of the fastener.

2. Place the fastener in position, being sure that each part is placed so that each sash will slide when the window is unlocked. Then partially engage the two parts of the fastener. While they are in this position, use the screw punch to make the screw pilot holes. Then screw each part to place.

3. Test the fastener by closing and locking the window. If the job has been done correctly, the fastener will hold the sash firmly together, eliminating all rattles.

*Installing Sash Lifts.* (See Figure 6.) The bar lift and the hook lift are centered and screwed to the face of the bottom rail of the lower sash. The flush lift requires mortising in, which is easily done by boring a series of holes about ¼ inch deep. The distance between the outer edges of the end holes is made a trifle greater than the dimensions of the protruding part of the lift. A sharp chisel will remove the wood between the holes. The bit selected must conform to the sash lift being fitted.

*Fitting Cabinet-door Hardware.* (See Figures 7 and 8.) Cabinet doors usually require two pieces of hardware: a wood, metal, or glass knob on the face of the door and a friction catch on the inside of the door to keep

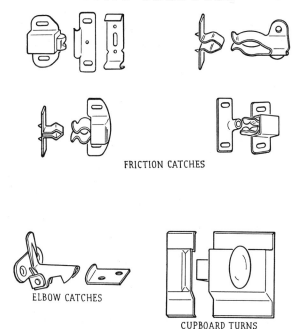

FRICTION CATCHES

ELBOW CATCHES

CUPBOARD TURNS

**Figure 7.** Cabinet hardware.

it closed. A cupboard turn serves as both a handle and a locking device.

Cabinet doorknobs are centered on the door stile and vertically placed slightly above the center of the door. A cupboard turn is placed flush with the outside edge of the door stile.

Friction catches or elbow catches are located on the underside of a cabinet shelf; the part screwed to the door is located so that it coincides exactly with the part screwed to the shelf. The catch will not work unless the two parts are lined up perfectly.

There are several types of friction catches available; none is difficult to fit; hence no detailed instructions will be given. Suffice it to say that the door part of the catch is screwed on first and that the keeper is then fastened to the *underside* of the shelf.

Wood doorknobs are fastened on by means of an oval-headed screw. Glass and metal knobs require small machine bolts. Bore a hole in the door stile slightly larger than the bolt. Be sure to work from both sides of the door or else to hold a small block of wood against the stile to prevent its splitting. A wood twist drill is best for boring. (Some twist drills are made for metal only.)

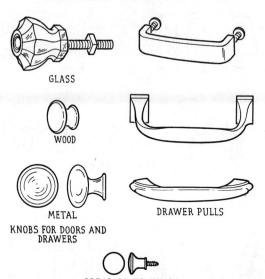

GLASS

WOOD

METAL          DRAWER PULLS

KNOBS FOR DOORS AND
DRAWERS

BREAD BOARD KNOBS

**Figure 8.** Cabinet knobs and pulls.

To summarize, proceed as follows when fitting cabinet-door hardware:
1. Select the hardware to be used. (Presumably it has been delivered to the job.)
2. Lay out the exact position of the doorknob.
3. Bore holes for the knob screws or bolts.
4. Screw on the knob.
5. Locate the position of the friction catch.
6. Screw the catch to the inside face of the door.
7. Screw the keeper to the underside of the cabinet shelf.
8. Close the door and check the installation.

***Fitting a Drawer Pull or Knob.*** (See Figure 8.) Drawer knobs are wood, metal, or glass. Drawer pulls are metal, either plated or solid brass. The metal and glass knobs are fastened on by means of a small bolt, which is screwed in from the inside of the front of the drawer; the wood knobs require ordinary wood screws. Metal hardware is applied to the face of the drawer front by means of wood screws or by small bolts, which are screwed through the drawer front. The bolts are purchased with the knob.

An ordinary drawer requires one pull or knob; wide drawers require two. When fitting the knobs or pulls to a series of drawers constructed one above the other, the hardware should be perfectly lined up.

When a single pull or knob is used, it is placed in the exact center of the width measurement of the drawer. It should be placed, vertically, slightly above the center. A drawer pull placed directly at center will appear to be below center.

The blueprints or specifications will state the type of pull to use. When metal knobs or pulls are used, they should match the hardware finish of the room.

If the knobs are wood, they are screwed on before the paint work is done; if metal or glass knobs or metal pulls are used, they are fitted *after* the painting is done. All holes for bolts should be bored prior to painting.

To fit a drawer pull, proceed as follows:

1. Select the correct hardware for the job. Specifications should be checked; presumably, on an inside-finish job, the hardware has already been delivered and is stored ready for use.

2. Mark the horizontal center of the drawer.

3. Mark a position slightly above the vertical center; the amount is optional—whatever looks good to the eye.

4. Using a wood twist drill, bore holes for the knobs, being sure to use a block on the inside face of the drawer to prevent splitting the wood. Then screw on the knobs.

5. If metal pulls are used, proceed as above: locate the exact position of the pull; make pilot screw holes; screw on the hardware, being sure to place it so that the screw slots are kept in the same position, that is, either vertical or horizontal, as desired.

***Installing Miscellaneous Hardware.*** (See Figure 9.) The term *miscellaneous* is used here to describe several pieces of finish hardware that are required as single items or that are classified as minor items. The list includes push plates, door grilles, coat and hat hooks, door bumpers, shutter knobs, and screen hangers.

PUSH PLATES. Push plates are used on double-acting doors, two being required for each door. The plates are made from polished glass, solid brass, or plated metal. The finish of the push plates, unless glass is used, should match the hardware in the room from which they can be seen.

The plates are centered on the door stile, with the center of each plate located approximately 42 inches from the floor. All screwheads should be left in a vertical position.

DOOR GRILLES. A door grille is used in a front door to provide a means of looking outside without opening the door. Grilles come in various sizes; $3\frac{1}{4}$ by 5 inches is standard, but larger grilles can be secured. Some are made with a door knocker attached.

The location of the grille depends on the design of the door. On a

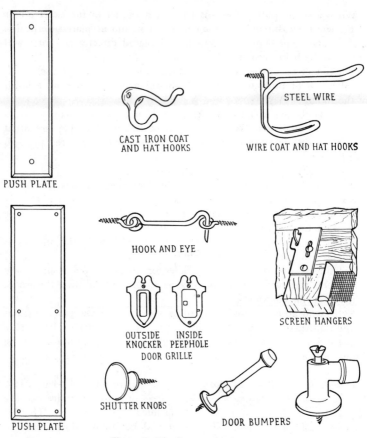

**Figure 9.** Miscellaneous hardware.

slab door the grille should be placed at eye height, as measured from the floor on the inside face of the door, unless the front doorstep is several inches below the sill line, in which case the grille should be a little lower. Horizontally, the grille is centered on the door. On a paneled door the location of the cross rails and vertical muntins determines the position of the grille.

To install a grille, proceed as follows:

1. Carefully lay out the exact position of the grille as determined by the door width and height measurements or by the door cross rails and vertical muntins. The size of the rectangular hole to be cut in the door should be slightly larger than the body of the grille. If the hole is made

too large, the outside edges of the grille will not cover it; if too small, the grille will not fit. The layout must be done on both sides of the door.

2. Bore a series of holes around the edges of the rectangle, being sure to work from both sides of the door; this is important; otherwise the wood bit will ruin the face of the door.

3. Remove the wood between the holes with a sharp chisel.

4. Place the grille in position and screw in the machine screws which clamp the inside part of the grille to the outside part.

COAT AND HAT HOOKS. Coat and hat hooks are of two general types, cast iron or wire; the former are much preferred. There is no standard spacing for the hooks, although they should be at least 8 inches o.c. The spacing is based on the lengths of the various pieces of hook strip.

For wire hooks, drill one pilot hole to make them turn in easily; for cast-iron hooks, drill holes for each screw.

DOOR BUMPERS. One door bumper is required for each communicating door to prevent the lock knob from defacing the plaster. Two types are made: one screws into the baseboard; the other is screwed into the floor. The wall bumper is preferable, but occasionally the placement of furniture necessitates the use of the floor type. Drill a pilot hole for each bumper and screw the bumper into place. The hardware finish of the bumper should match the finish of the room.

SHUTTER KNOBS. Shutter knobs are used on interior louvers and exterior shutters. The knobs should be centered on the shutter stiles. Bore a pilot hole and screw in the knob. Wood is used for exterior knobs and painted to match the shutter; interior knobs should match the room hardware.

SCREEN HANGERS. Each window screen requires one pair of screen hangers, located as shown in Figure 9. First, screw the hook to the edge of the blind stop, being sure that the hook projects beyond the face of the stop the correct amount to receive the slot in the flat piece of the hanger. Then screw in the top screw that holds the flat piece, centering the screw in the screw slot which permits vertical adjustment; next, hang the screen in place and test it for ease of operation, making sure that it does not bind; then put in the second screw.

One hook and eye is used at the bottom of the screen to hold it closed. First, the hook is centered, horizontally, on the bottom rail. and then the screw eye is placed to permit the hook to enter it.

*Installing Toggle or Molly Bolts.* (See Figures 10 and 11.) In modern houses, wood lath has been replaced by various forms of plasterboard. In some instances, plaster is applied to such boards. In other cases, as in dry-built construction, the boards form the complete wall finish. Plasterboard will not hold the screws required for the appli-

**Figure 10.** Toggle bolt and Molly bolt.

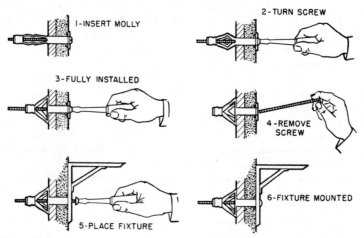

**Figure 11.** Steps in installing a Molly bolt after hole is bored. (Courtesy of Molly Corporation.)

cation of drapery cranes, curtain rods, towel bars, etc., unless wood backing has been provided. Where no backing is provided, a special fastener, either a toggle bolt or a Molly bolt, must be used.

Both toggle and Molly bolts work on the principle of expanding fingers. To use either, a hole must be drilled through the plasterboard. The bolt is inserted into the hole and turned by means of a screwdriver. This action expands the fingers on the inside face of the plasterboard so that they hold the bolt securely in place. Any fixture can be attached with either type of bolt.

**Note:** Molly bolts can be purchased in various sizes, ranging from ¼ inch to ⁷⁄₁₆ inch in diameter and from 1⅜ inches to 3½ inches in over-all length. Since Molly bolts are widely used, the five steps required to install a bracket by this means are illustrated in Figure 11.

# Index

# BLUEPRINTS
*Plates I to VI*

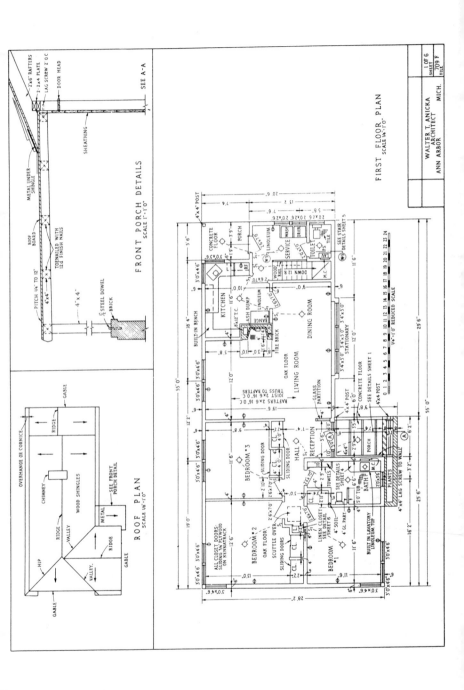

FRONT PORCH DETAILS
SCALE 1"=1'0"

2×6 RAFTERS
2×4 PLATE
LAG SCREW 2' OC
DOOR HEAD

SEE A-A

METAL UNDER SHINGLE

SHEATHING

PITCH 1/4" TO 12"

TOENAILD WITH
12d FINISH NAILS

ROOF BOARD

4×4

4"×4"

STEEL DOWEL

BRICK

ROOF PLAN
SCALE 1/8"=1'0"

OVERHANGE OR CORNICE

GABLE
RIDGE
CHIMNEY
RIDGE
WOOD SHINGLES
SEE FRONT PORCH DETAIL
METAL
HIP
RIDGE
VALLEY
VALLEY
VALLEY
GABLE
GABLE

FIRST FLOOR PLAN
SCALE 1/4"=1'0"

KITCHEN
PORCH
CONCRETE FLOOR
SERVICE
WASH  DRYER
TOILET
DOWN 12
M.C.
LINOLEUM
DINING ROOM
STATIONARY 12'0"
FIRE BRICK
LIVING ROOM
OAK FLOOR
GLASS PARTITION
4×4 POST
CONCRETE FLOOR
PORCH
SEE DETAILS SHEET 1
BUILT-IN BENCH
ASH DUMP
RANGE
RAFTERS 2×6 16"O.C.
JOIST 2×6 16"O.C.
TRUSS TO RAFTERS
BEDROOM 3
SLIDING DOOR
CL.
CL.
HALL
RECEPTION
SLIDING DOOR
CL.
BEDROOM 2
OAK FLOOR
SCUTTLE OVER
SLIDING DOORS
CL.
CL.
LINEN CLOSET
SEE DETAIL SHEET 6
4" SOIL
BEDROOM 1
BUILT IN LAVATORY
LINOLEUM TOP
TOWELS
BATH
PLANTS
4×4 LAG SCREW TO WALL
4×4 POST
ALL CLOSET DOORS:
SLIDING 3/4" PLYWOOD
ON KENNATRACK
SEE DETAILS SHEET 4

WALTER T. ANICKA
ARCHITECT
ANN ARBOR   MICH.

SHEET 1 OF 6
FILE 709 F

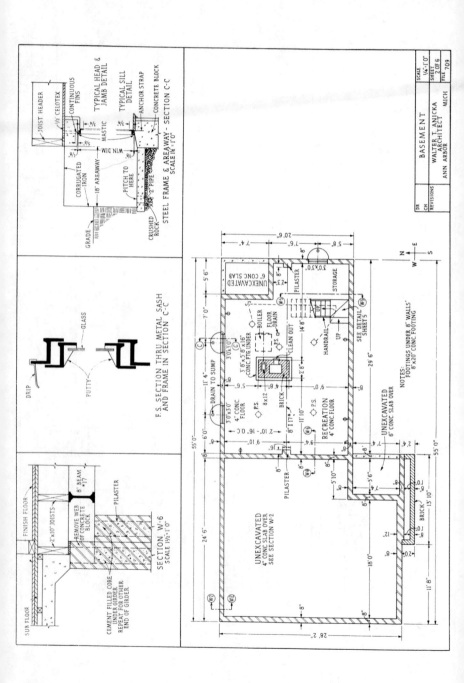

BASEMENT

WALTER T. ANICKA
ARCHITECT
ANN ARBOR MICH

SCALE ¼"=1'0"
SHEET 2 OF 6
FILE 709

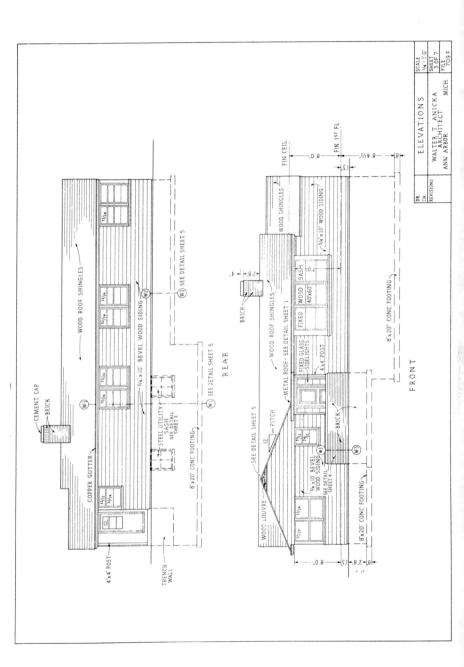

REAR

- CEMENT CAP
- BRICK
- WOOD ROOF SHINGLES
- COPPER GUTTER
- 4"x4" POST
- TRENCH WALL
- STEEL UTILITY SASH SEE DETAIL SHEET 2
- ¾" x 10" BEVEL WOOD SIDING
- 8"x20" CONC FOOTING
- (W₂) SEE DETAIL SHEET 5
- (W₃) SEE DETAIL SHEET 5
- (W₁)
- (W₂)
- GL

FRONT

- FIN CEIL.
- FIN 1ST FL.
- WOOD SHINGLES
- ¾" x 10" WOOD SIDING
- 8"x20" CONC FOOTING
- BRICK
- WOOD ROOF SHINGLES
- METAL ROOF-SEE DETAIL SHEET 1
- FIXED GLASS SIDELIGHTS
- FIXED WOOD SASH 40"x60"
- 4"x4" POST
- SEE DETAIL SHEET 5
- 12 PITCH
- ¾"x10" BEVEL WOOD SIDING SEE DETAIL SHEET 4
- WOOD LOUVRE
- 8"x20" CONC FOOTING
- BRICK
- (W₁)
- (W₃)

ELEVATIONS
WALTER T. ANICKA
ARCHITECT
ANN ARBOR    MICH.

| DR | | SCALE ¼"=1'0" |
| CH | | SHEET 3 OF 7 |
| REVISIONS | | FILE 709 F |

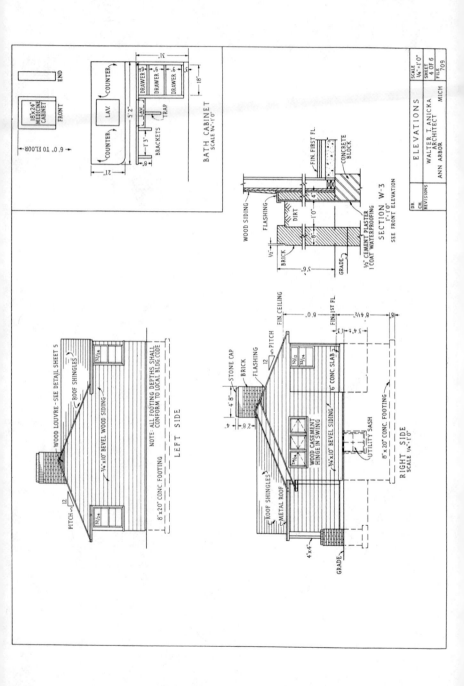

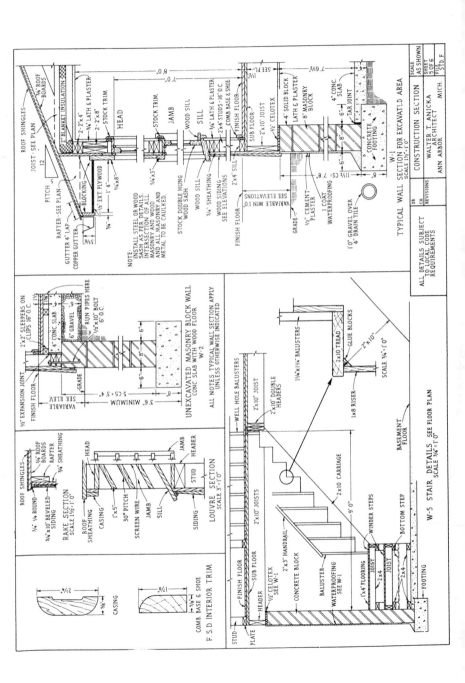

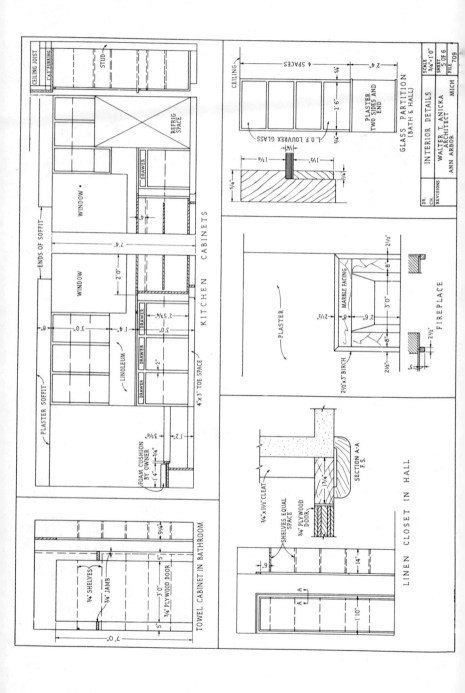

KITCHEN CABINETS

CEILING JOIST
1" x 3" FURRING
STUD

WINDOW

REFRIG SPACE

DRAWER

ENDS OF SOFFIT

PLASTER SOFFIT

WINDOW

LINOLEUM

DRAWER 2½'-5¾"
DRAWER 2"
DRAWER

3'-0"
1'-4"
3'-0"
2'-0"
7'-4"

8"

4'' x 3'' TOE SPACE

FOAM CUSHION BY OWNER

3⅝"
1'-2"
1'-4"
¾"

TOWEL CABINET IN BATHROOM

9¾"

¾" SHELVES
¾" JAMB
¾" PLYWOOD DOOR

5"
3'-0"
5"
7'-0"

GLASS PARTITION
(BATH & HALL)

CEILING

4 SPACES

2'-6"
¾"
¾"
2'-4"

L.O.F. LOUVREX GLASS

PLASTER TWO SIDES AND END

1-¾"
¼"
1-¾"
5¼"
¼"

FIREPLACE

PLASTER

MARBLE FACING

2½"x3" BIRCH

2½"
8"
3'-0"
8"
2½"
2½"

6"
2'-6"
21½"/½"
6"

LINEN CLOSET IN HALL

¾" x 1½" CLEAT

SHELVES EQUAL SPACE

¾" PLYWOOD DOOR

SECTION A·A
F.S.

1¾"

6"

14"

A
A

1'-10"

INTERIOR DETAILS

SCALE
¾"·1'-0"

WALTER T. ANICKA
ARCHITECT
ANN ARBOR        MICH

DR.
CH.

REVISIONS

SHEET
5 OF 6
FILE 709

Catalog

If you are interested in a list of fine Paperback
books, covering a wide range of subjects
and interests, send your name and address,
requesting your free catalog, to:

McGraw-Hill Paperbacks
330 West 42nd Street
New York, New York 10036